Switches

Switches

Ian R. Sinclair

BSP PROFESSIONAL BOOKS

OXFORD LONDON EDINBURGH

BOSTON PALO ALTO MELBOURNE

First published 1988

British Library
Cataloguing in Publication Data
Sinclair, Ian R.
Switches
1. Switching circuits
I. Title
621.3815'37 TK7868.S9

ISBN 0-632-02068-7

BSP Professional Books
A division of Blackwell Scientific Publications Ltd
Editorial Offices:
Osney Mead, Oxford OX2 0EL
(Orders: Tel. 0865 240201)
8 John Street, London WC1N 2ES
23 Ainslie Place, Edinburgh EH3 6AJ
3 Cambridge Center, Suite 208, Cambridge, MA 02142, USA
667 Lytton Avenue, Palo Alto, California 94301, USA
107 Barry Street, Carlton, Victoria 3053, Australia

Set by DP Photosetting, Aylesbury, Bucks
Printed and bound in Great Britain by Mackays of Chatham, Kent

Contents

Preface vii
1 Principles 1
2 Non-signal Switches 17
3 Signal-carrying Switches 33
4 Other Mechanisms 57
5 Specifications and Standards 91
Index 101

Preface

Most texts and courses in electronics take switches and switching action completely for granted, so that this essential electronics component is quite remarkably neglected. Virtually every electronic device makes use of at least one switch, which has to be specified by the design engineer and maintained or replaced by the service engineer or technician. All of them need information on this component just as much as for any other device in a circuit. The vast variety of switch types, as much as their essential function, is a good reason for providing some information about these components, but even more important is the need to show what the switch does and why there is any choice in the specification. These are topics that are generally very poorly covered in courses dealing with electronic components.

This book does not cover every possible variety of switch, because to do so would be a mammoth task. The switches that are dealt with here are the low-voltage and (comparatively) low-current switches that are used mainly in electronics circuit applications. The voltage range of up to 250 V and the current range of up to 20 A have been selected arbitrarily as the cut-off points as far as the range of supply-voltage switches is concerned. More importantly, however, the book covers the other waveforms that switches can be expected to deal with: the audio, video, digital and VHF/UHF switching requirements that are so often met with nowadays.

The understanding of the need for a choice of contact materials is a topic that requires some background in chemistry and physics. In this book, the background information will be part of the text, as I cannot assume that everyone who needs to know the contact metal requirements for a switch will necessarily have the knowledge of these subjects that is so often assumed. The book has been written to be as useful as possible to a readership whose interests are not specialised, other than in some branch of electronics. Because of this, the book should be of considerable help to electronics engineers, students, technicians, and anyone who is required to specify electronic components. A section on legislation and standards has been written with a view to assisting in the selection of switches for applications where choice can be restricted because of legal requirements, or because work is being done to military or other standards.

I am most grateful to the IEE and to the BSI for permission to quote from various publications, and to RS Components for data on many switch types. I am also indebted to Bernard Watson of Blackwell Scientific Publications Ltd for guidance and encouragement, and, again, to RS Components for commenting on the manuscript.

Ian R. Sinclair, Autumn 1987

Chapter One
Principles

Switches are almost as old as present-day electricity applications, which is probably why their action is taken so much for granted. The function of a switch is to make or break current in a circuit, and in the early nineteenth century this would have meant only a DC circuit. Later, switching had to be applied to AC, later still to audio and radio frequencies and video signals, and most recently to digital signals. Early switches were simple mechanical devices, the knife switches that can still be seen in old factories. This century has seen the development of much more elaborate mechanical principles, and of switches with no mechanical moving parts at all. In this book, however, we are mainly concerned with mechanically-operated switches, and though some electronic types are dealt with, relays and contactors have been excluded.

The switch, then, is a device with a circuit path that can be made or broken. When the circuit path is made, the circuit resistance through the switch is low, of the order of milliohms for most switches. With the circuit path broken, the circuit path through the switch has a high resistance, several M or higher. In addition, the resistance between the switch circuit path and the body of the switch, including the operating mechanism and the mountings, is generally required to be high. This resistance, and the maximum voltage that can be applied to this insulation, is often of major importance in mains voltage switches, and is a vital feature of legislation regarding switch suitability. For some applications to high-frequency signals the capacitance between the switch circuits and the switch body is of considerable importance, and will decide the design of the switch.

Contact resistance

The resistance of the (mechanical) switch circuit when the switch is on (made) is determined by the *switch contacts*, the moving metal parts in each part of the circuit which will touch when the switch is on. The amount of the contact resistance depends on the area of contact, the contact material, the amount of force that presses the contacts together, and also the way that this force has been applied. For example, if the contacts are scraped against each other (a *wiping* action) as they are forced together, then the contact resistance can often be much lower than can be achieved when the same force is used simply to push the contacts straight together. In general, large contact areas are used only for high-current operation, and the contact areas for low-current switches (as used for electronics circuits) will be small. The actual area of electrical connection will not be the same as the physical area of the contacts, because it is generally not possible to construct contacts that are precisely flat, or have surfaces that are perfectly parallel when the contacts come together. This problem will be familiar to anyone who has renewed the ignition points in a car in the days before electronic ignition circuits. The actual area of contact of the switch is revealed by the contact burning when the switch is used in inductive circuits, but is not necessarily obvious in other cases. Since the size of the switch to a large extent determines the amount of contact pressure that can be used, and the area of contact is rather indeterminate, the main factor that affects contact resistance is the material used to make the contacts themselves.

Wiping contacts

One method of reducing contact resistance in difficult conditions is to use a *wiping* action. As the name suggests, the mechanical action of the switch is such that the contacts are rubbed against each other as they make connection. This will normally remove any thin films of oxide or other contaminants, and so ensure a much lower contact resistance than could be obtained by bringing the contacts together in a more straightforward manner. The disadvantage of a wiping action is that it can abrade the contacts, and if these consist of steel or nickel-alloy plated with gold, the gold plating can be removed by this abrasion. Wiping contacts are best suited to switches that are infrequently used and located in contaminating environments, and those in which silver is used as the contact material.

The switch designer has the choice of making the whole of a contact from one material, or of using electroplating to deposit a more suitable contact material. By using electroplating, the bulk of the contact can be made from any material that is *mechanically* suitable, and the plated coating will provide the material whose resistivity and chemical action is more suitable to the *electrical* function of the switch. Plating also makes it possible to use materials such as gold and platinum, which would make the switch impossibly expensive if used as the bulk material for the contacts. It is normal, then, to find that contacts for switches are constructed from steel or from nickel alloys, with a coating of material that will supply the necessary electrical and chemical properties for the contact area. The choice is never easy, because the materials whose contact resistance is lowest are in general these which are most vulnerable to chemical attack by the atmosphere. In addition, some materials which can provide an acceptably low contact resistance exhibit 'sticking', so that the contacts do not part readily and may weld shut. The other main problems are burning and oxidation. The spark current which passes at the time when contacts are opening can induce melting of the contacts, or cause the metal to combine chemically with the oxygen in the atmosphere (oxidation or 'burning'). Of these two, oxidation is the more severe problem, because the oxides of most metals are non-conductors. This means that oxidation causes a large rise in contact resistance, even to the point of making the switch useless.

The usual choice of materials is illustrated in Figure 1.1. From the point of view of resistance alone, silver is the preferred material, since silver has the lowest resistivity of all metals. Unfortunately, silver is very badly corroded by the atmosphere, particularly if any traces of sulphur dioxide exist (where coal or oil is burned, for example), and silver contacts will have a short life if there is any arcing at the contacts (see below). Silver contacts are desirable if the lowest contact resistance is needed for high-current use, but the action of the switch will have to be such that the contact pressure is high, and the contacts are wiped as they are brought together. Silver-coated contacts can be used with fewer problems when the switch is sealed in such a way as to exclude atmospheric contamination, or if the contacts are surrounded by inert gases, but these are not simple solutions to the corrosion problem.

Gold plating is a very common solution to the contact problem,

Material	Advantages	Disadvantages	Uses
Silver	low resistance	corrodes	high current, high contact pressure
Palladium-silver	less easily contaminated	higher resistance	general use
Silver-nickel	resists burning and sticking	higher resistance	general use
Tungsten	hard, high melting point	oxidises easily	high-power switching
Platinum	resists chemical attack, stable	high voltage, low current use only	specialised
Gold	resists corrosion, can be plated onto other metals	low current only	widespread in electronics

Fig. 1.1 Comparison of commonly-used contact materials for switches and relays.

particularly in switches for low-current electronics use. The contact resistance can be moderately low, and the gold film is soft, so that moderate pressure can result in a comparatively large area of contact. Of all metals, gold is about the most resistant to corrosion, though the combination of hydrochloric acid and electric current can cause gold to be attacked quite rapidly, making this material unsuitable if the atmosphere contains traces of chlorine or hydrochloric acid. Since passing electric current through sea-water causes chlorine to be generated, gold-plated contacts can be severely corroded when the switches are used at sea. The switches are suitable only for the lower currents, but since most switch applications in electronics *are* for low currents this is no handicap. Gold plating of contacts is often mandatory in the specification of switches for military contracts.

Contact coatings based on the 'noble' metals are often employed. These metals are so named because, like gold, they exhibit a high resistance to chemical attack; typical metals in this group are platinum, palladium, iridium and rhodium. All of them have rather high contact resistance, but are very stable and resist chemical attack. Platinum is particularly useful for contacts that will be used for low currents and for high voltage levels. Rhodium and iridium platings provide a high level of resistance to corrosion along with stable contact resistance, and are suitable for medium-voltage, medium-current applications. Some advantages can be gained by using thick films of contact materials which are alloys of the noble metals with silver. For example, palladium-silver has a much better resistance to contamination than silver itself, though with a higher contact resistance than silver. It is one of the general contact materials that can be used in switches for various types of application.

The metals tungsten and molybdenum, though not of the platinum group, are also used as contact materials for special purposes. Tungsten in particular is very resistant to burning caused by contact arcing, and is used for high-power switching. Its disadvantage is that the surface will oxidize easily, causing contact-resistance problems.

The most common general-purpose contact alloy material, however, is nickel-silver. The cost of nickel-silver is such that it can be used as a bulk material rather than as a coating, and though its contact resistance is higher than that of pure silver, it is much more resistant to chemical attack than the pure metal. It also resists burning, and the contacts do not tend to stick together.

AC and DC switching

The type of current that is being switched has a very considerable effect on the life of the switch contacts. If we confine ourselves to currents that are not signals, then the switch may have to handle either AC or DC. When contacts are made, the type of current is almost immaterial, because there is very little sparking across contacts that are rapidly coming together. Any sparking or arcing takes place almost exclusively when contacts are being separated. This is because conditions for sparking or arcing are ideal at the instant when two contacts are separating. At this instant, the gap between the contacts is very small, so that only a small voltage difference is needed to make current cross the gap. In this respect, *sparking* means an intermittent discharge between the contacts, and *arcing* is a continuous discharge.

Arcing is a much more damaging effect, because very high temperatures can be generated locally at the points where the arc originates. In addition, the air itself between the contacts will be heated to very high temperatures, and this has the effect of ionizing the air, making it conduct and so encouraging the arcing, which will then continue even when the contacts are separated by considerably more than the normal sparking distance.

A DC supply is much more likely to cause arcing than an AC supply, given the same conditions of contact opening speed. This is because the voltage across the opening contacts will remain fairly steady, whereas for an AC supply the voltage will drop to zero in each half-cycle. If the opening time of a switch used for AC is reasonably short, then the contacts will have opened appreciably while the voltage across them is, on average, low. In addition, the effect of metal transfer (see below) will be more serious when DC is switched. This difference is reflected in the very different ratings that apply for AC and for DC use of any given switch. One example is a miniature mains rocker switch rated at 250 V AC 4 A and 28 V DC 4 A, so that the effect of using DC is effectively to derate the switch operating voltage by a factor of nine times. This is not at all unusual, though the DC voltage rating might be increased if the current rating were decreased.

Inductive loads

Whether the supply that is being switched is AC or DC, the presence of an inductor of appreciable size in the switched circuit has a large effect on the switch ratings. Once again, the presence of an inductor (of the order of 0.05 H upward) when a circuit is made presents no difficulties. The effect of an inductor on making a circuit is to cause the current in the circuit to rise to its normal value more slowly than for a resistive circuit. The problems arise when an inductive circuit is broken. When current through an inductor is decreased, a voltage is induced across the inductor, and the size of this voltage is equal to inductance multiplied by rate of change of current. The size of the inductance will be expressed in Henrys and the rate of change of current in amps per second.

Suppose, for example, that a circuit contains a 0.5 H inductor, and a switch would break a current of 2 A in a time of 10 ms (0.01 s). The rate of change of current here is $2 \div 0.01 = 200$ amps per second, so that the voltage across the inductor is $0.5 \times 200 = 100$ V. Now this is comparatively small, but if the circuit happened to be a 6 V DC circuit, with the switch contacts rated for 6 V DC, then the presence of 100 V across the switch contacts will cause a considerably greater amount of arcing than would be present in a resistive circuit. In practice the effective breaking time is likely to be less than 10 ms, so that the voltage induced by breaking the circuit will be higher than this example suggests. If the circuit uses AC, then there is a reasonable chance that the current may not be at its peak at the time when the circuit is broken and, as before, the amount of arcing is likely to be less. Note, however, that the induced voltage across the inductor at the instant when the circuit is broken is not an AC voltage, and follows much the same pattern whether the current that flowed was AC or DC. The effect of breaking the inductive circuit is a pulse of voltage, and the peak of the pulse can be very large, so that arcing is almost certain when an inductive circuit is broken unless some form of arc suppression is used.

Arcing and switch life

Arcing is one of the most serious of the effects that reduce the life of a switch. During the time of an arc, as we have seen, very high temperatures can be reached, both in the air and on the metal of the contacts. The ionization of the air gives rise to a *plasma*, a quantity of completely ionized air which is a comparatively good conductor of electric current. The temperatures that are reached in this plasma can cause the metal of the contacts to vaporize and be carried from one contact to the other. This effect is very much more serious when the contacts carry DC, because the metal vapour will also be ionized, and the charged particles will always be carried in one direction. If, for example, the metal forms positive ions, then it will always be carried from the positive contact of the switch to the negative contact. This will result in the familiar effect of the positive contact developing a crater and the negative contact developing a mound. Since the surfaces of these parts of the contacts are rough, this greatly reduces the contact area, lessening the ability of the contacts to take their rated current. As usual, AC causes less trouble, because the arc is usually quenched within a half-cycle, and since the current direction will not always be the same at the time when the arc exists, the transfer of metal is not always in one direction.

Arcing is also a means by which contacts become contaminated, because in the plasma that exists during arcing any contaminant elements in the atmosphere become ionized, and are transferred to one of the contacts. By this means contaminants that were present in the atmosphere become permanently embedded in the contacts, with detrimental effects on the contact resistance. Arcing can also cause severe oxidation of the metal of the contacts, and this will also raise contact resistance. Fortunately the transfer of metal in arcing will inhibit oxidation to some extent, because it causes a fresh set of metal surfaces to appear at each switching-off action.

Arcing, therefore, will greatly reduce the life of switch contacts, and it is an effect that cannot be completely eliminated. Arcing is almost imperceptible if the circuits that are being switched run at low voltage and contain no inductors, because a comparatively high voltage is needed to start an arc. For this reason, then, arcing is not a significant problem for switches that control low voltage, such as the 5 V or 9 V DC that is used as a supply for solid-state circuitry, with no appreciable inductance in the circuit. Even low-voltage circuits, however, will present arcing problems if they contain inductive components, and these include relays and electric motors as well as chokes. Circuits in which voltages above about 50 V are switched are the most susceptible to arcing problems, particularly if inductive components are present, and some consideration should be given to selecting suitably rated switches, and to arc suppression, if appropriate.

Arc suppression

The total elimination of arcing is never possible, though arcing becomes insignificant if the current carried by a switch is low (of the order of a milliamp), and with a non-inductive load. If a mechanical switch is to have a very long life, consideration should be given to using it in the input side of a solid-state switching arrangement, using transistors (bipolar or MOS), thyristors or Triacs. This solution may not be appropriate to every type of switching application, but for the switching voltages and currents that are used in electronics work it can often be a way out of a reliability problem. One snag is that the use of a mechanical switch along with an electronic switch requires a power supply to the semiconductor part of the circuit, which will need to be applied before the mechanical switch can be operated. This may mean that two switches are needed: the second applies power to the circuitry that will then switch the main load. If the switches are then used in the wrong order, the arcing problem will return. For very large switchgear, solutions such as oil-quenching and air-blast arc suppression are used, but in this book we are confining our attention to the smaller size of switch, and hence to purely electrical and electronic methods.

Most arc-suppression circuits hinge on delaying the change of voltage across the switch contacts, and the techniques that are used

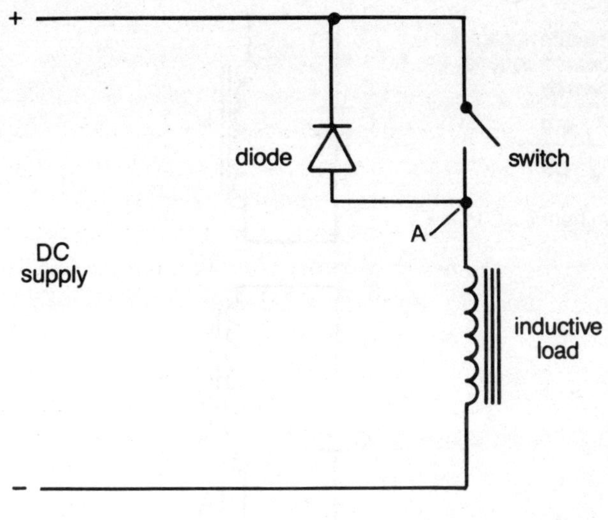

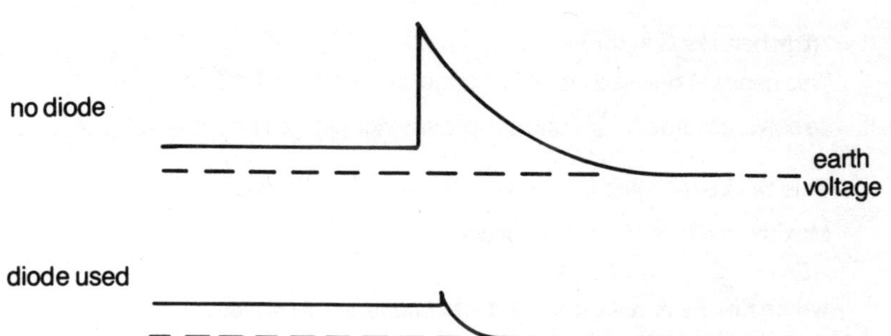

Fig. 1.2 Diode transient suppression connected across switch contacts in an inductive circuit.

depend on the type of supply (AC or DC) and the nature of the load (inductive or non-inductive). Since DC supplies and inductive loads present the main problem, we shall deal with these first. The most effective arc suppression for circuits of this type is diode suppression, as illustrated in Figure 1.2. The principle is that when the circuit is broken, the sudden drop of current through the inductive load will induce a voltage which is always in the reverse direction to the applied voltage across the switch. By using a diode connected so as to conduct for this polarity of voltage, the surge can be forced to pass a pulse of current back to the power supply, or into a capacitor wired between the switch live terminal and earth. This form of protection is particularly

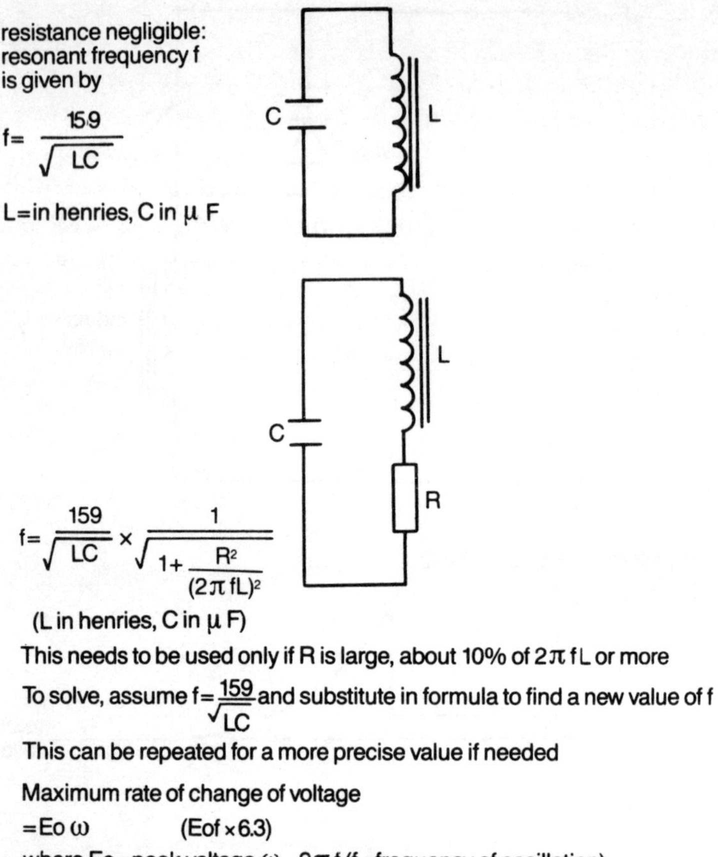

Fig. 1.3 Resonance in capacitor-inductor circuits. The resistance in a switch suppression circuit is usually large, but the precise frequency of resonance is not important in any case.

effective when the supply voltage is low, because it protects the switch against the much higher voltages which are caused by breaking the inductive load, and which are high enough to cause arcing. Diode protection is much less effective when the circuit operates at a higher voltage, 100 V or more, because the voltage across the contacts cannot be less than the supply voltage, and this is large enough to promote arcing. The arcing will, however, be considerably less severe than it would be if the diode were not present. Whatever voltage level is used in the circuit, the diode must be adequately rated both for the voltage and for the current that can flow. The reverse voltage rating of the diode should be adequate for the circuit applied voltage, and preferably higher so as to give a margin of safety. Note that in such a circuit, if the diode were to break down to a short-circuit the circuit would be switched on, and this could affect safety. No form of arc-suppression diode or other circuit should be applied across any form of safety switch. For such switches, arc suppression is unimportant because the switches should not normally be opened with power applied, and the risk of failure of a suppression component might have disastrous results.

The other main electronic method of arc suppression in DC circuits involves the use of capacitors, often along with resistors and/or inductors. The use of a capacitor alone relies on the rate of charge of the capacitor being lower than the rate of change of voltage across the switch when it is opened. This will certainly be true for a non-inductive circuit, and will be true for an inductive circuit if the resonant frequency of the capacitor with the inductive load is low. Figure 1.3 gives a guide to resonant frequency and the relationship between rate of change of voltage and frequency. For many purposes a resonant frequency of 50 Hz is a good choice, because the AC rating of the switch is then applicable to its use in the DC arc-suppressed circuit. The capacitor that is used must be rated for the peak voltage that can be expected, and must be of the paper or plastic dielectric type. Never specify an electrolytic capacitor for this type of action, though ceramic or mica types can be used additionally as suppressors for high-frequency oscillations if needed. Note also that this capacitor method of arc suppression, using comparatively large values of capacitors, applies to DC circuits. For an AC supply, the presence of a capacitor will provide a leakage path when the switch is open-circuit, and this can cause problems, not least with safety. Any suppression components in an AC supply should use capacitors only across the lines (Figure 1.4), never across the switch contacts themselves, though RF suppression capacitors of a few pF are permissible in some circuits.

The use of capacitors with resistors and inductors leads to more elaborate suppression circuits. A series capacitor-inductor suppression circuit will have a series resonance at which its impedance is a minimum, and a rate of change of voltage corresponding to this resonant frequency will pass easily through the suppression circuit in preference to the (open) switch contacts. This type of suppression is particularly useful if diodes cannot be used. By adding a resistor, either to the capacitor-inductor circuit or along with the capacitor only, the

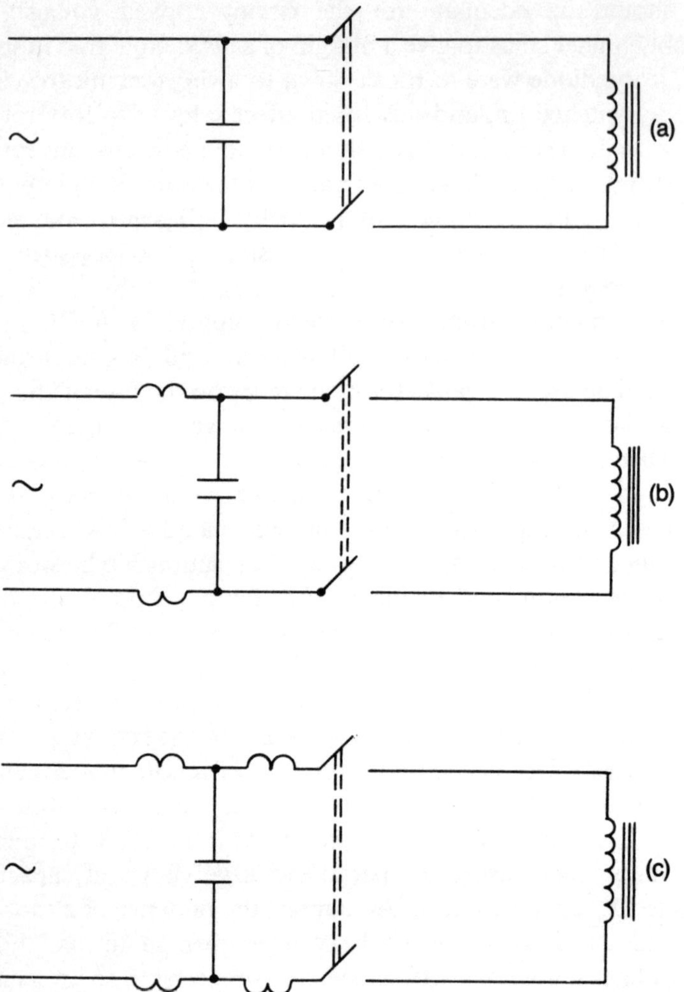

Fig. 1.4 Circuits used on AC supply lines for the suppression of transients caused by switching. This can include the interference caused by the use of thyristors and Triacs.

rate of change of current through the arc-suppression circuit can be reduced. This in itself does not assist in arc-reduction – quite the reverse – but it can greatly reduce radio-frequency interference (RFI) caused by the arc-suppression circuit itself. This in turn should be rather less than the RFI that would be caused by the arcing. Figure 1.5

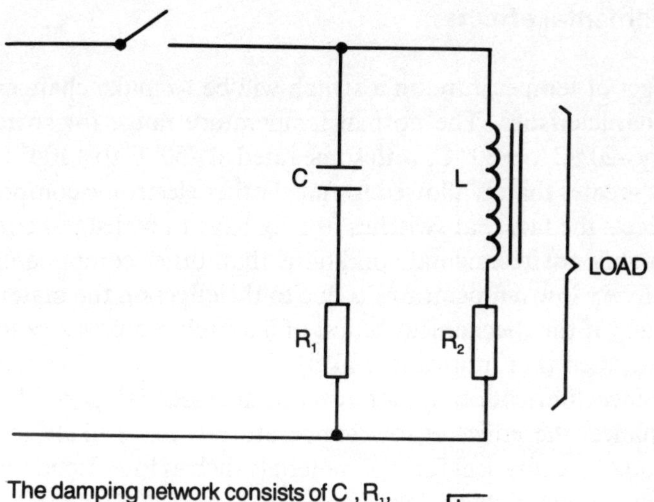

The damping network consists of C, R_1, damping is maximum when $R_1 = R_2$ and $R_1 = \sqrt{\frac{L}{C}}$ so that the requirements are:

$$R_1 = R_2$$

$$C = \frac{L}{R_1^2}$$

C in farads, L in henries, R in ohms

Fig. 1.5 Using a capacitor-resistor damping network connected across switch contacts.

shows circuits and formulae relating to resistor-capacitor-inductor suppression circuits.

For AC supplies, the nature of the supply voltage itself ensures that any arcing should be quenched in a fraction of a cycle, so that the AC current rating for switches is generally higher than the DC rating. The use of diode arc suppression is ruled out, and in general capacitors and inductors are used only in order to suppress RFI, using mica capacitor and ferrite-cored inductors. Another method of suppression is the provision of non-linear resistors across the switch contacts. The type of non-linear resistor that can be used has a voltage-current characteristic that shows a decrease of resistance as the voltage across the resistor increases. In many applications, however, the polarity of any excessive voltage across a switch can be predicted, so that a semiconductor diode can be used for absorbing any surges.

Environmental effects

The effect of temperature on a switch will be to make changes in its rated characteristics. The normal temperature range for switches is typically –20° C to +80° C, with some rated at –50° C to +100° C. This range is greater than is allowed for most other electronic components, and reflects the fact that switches usually have to withstand considerably harsher environmental conditions than other components. The effect of very low temperatures is due to the effect on the materials of the switch. If the mechanical action of a switch requires any form of lubricant, then that lubricant is likely to freeze at very low temperatures. Since lubrication is not usually an essential part of switch maintenance, the effect of low temperature is more likely to be an alteration of the physical form of materials such as low-friction plastics and even contact metals.

Plastics that are used as low-friction bearings can become sticky at very low temperatures, causing irregular mechanical action. The metals that are used for contacts can alter their crystal shape, causing a large increase in contact resistance. This change of crystal shape is very sharp, and will occur at a clearly-defined temperature. The best-known change of this type is the change of tin from its metallic form to the form of a dull-grey powder at a temperature of about –20° C, but other metals will exhibit less dramatic changes at various temperatures. At the high-temperature end of the scale, the most serious threat is to insulation resistance, which can decrease very considerably at the higher temperatures. Contact resistance can also be increased because of the growth of layers of oxide on contact surfaces, and if a switch is to be used at constant high temperatures, near its rated limit, a wiping action is desirable (see above). For electronics purposes, exposure of a switch to high temperatures is more likely than exposure to very low temperatures.

Sealing and flameproofing

The flameproofing of a switch does not refer to protection of a switch against fire, but to the prevention of fire caused by the switch. A flameproof switch would be specified wherever flammable gas can exist in the environment (for instance in mines, chemical stores, and processing plants that make use of flammable solvents). In such an atmosphere sparking or arcing at switch contacts could ignite flammable gas in the atmosphere, with potentially devastating results. A flameproof switch is one which is sealed in such a way that sparking at the contacts can have no effect on the atmosphere outside the switch. This implies that no vapour from the atmosphere can penetrate the switch in any way, either through cable entry points, or where the switch actuating mechanism enters the switch. This makes the preferred type of mechanism the push-on, push-off type, since the push button can have a small movement and can be completely encased along with the rest of the switch. Flameproofing cannot necessarily be satisfactorily achieved simply by sealing the switch, because a sealed switch can often have a very short life because of the accumulation of acids caused by the effect of sparking and arcing on the confined atmosphere. A flameproof switch requires materials that are selected so as to minimise attack by such acids, and a sealing material that will not contribute unwanted vapours into the switch. For preference it should be filled with an inert gas such as nitrogen. The materials used for sealing nowadays are usually synthetic silicone rubbers, because these are themselves very inert, and can also withstand high temperatures without deterioration. Since flameproof switches are intended for situations where safety is of paramount importance, all such switches must conform to the appropriate British Standard BS 6458.

The sealing of a switch may also be intended to provide waterproofing, and this is covered by the BS 2011 set of tests, Parts 2.1 Q and R. A major problem of waterproofing is that the switch becomes non-vented, so that any arcing will result in a corrosive atmosphere unless the switch is surrounded by inert gas before sealing. Any switch that is part of a circuit likely to work for long periods immersed should preferably handle low voltage and current with a non-inductive load. This is not difficult to arrange if the switch is part of a system for which the main circuits are in air.

Connections to switches

Connections to switches may be made by soldering, welding, crimping or by various connectors or other plug-in fittings. The use of soldering is now comparatively rare, because unless the switch is mounted on a PCB which can be dip-soldered this will require manual assembly at this point. Welded connections are used where robot welders are employed for other connection work, or where military assembly standards insist on the greater reliability of welding. By far the most common connection method for panel switches (as distinct from PCB-mounted switches) is crimping, because this is very much better adapted for production use. Where printed circuit boards are prepared with leads for fitting into various housings, the leads will often be fitted with bullet or blade crimped-on connectors so that switch connections can be made. This assumes the use of switches with suitable terminations, and the most common switch termination of this type is the flat blade, so the connecting cable should carry a matching receptacle. Piggy-back receptacles can be used when wiring has to be carried from one switch to another.

Chapter Two
Non-signal Switches

Requirements

The category of non-signal switches includes all varieties of switch that are placed in supply voltage lines, whether DC or AC, or which make and break contacts between DC levels that are not of critical amplitudes. Switches that are used in power supply leads will often have a mechanical action that provides for a rapid switch-over, the snap-over type of mechanism that is illustrated in Figure 2.1. This

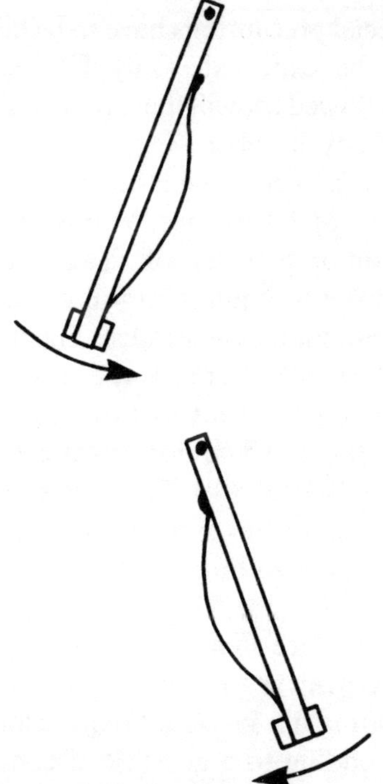

Fig. 2.1 The principle of the snap-over action of switches. The mechanism illustrated here is used on microswitches.

provides a form of mechanical hysteresis, so that the position of the switch operating mechanism for making connection is not the same as the position for breaking connection. This snap-over action is not provided on all types of mechanical switches, however, and may be undesirable if a slow opening rate is required. In this chapter we shall confine our attention to manually operated switches, because any discussion of relays and contactors would extend the scope of the book too far into more specialised topics.

The simplest switch requirement to fulfil is for high-voltage (in the region of a few hundred volts), low-current, non-inductive load, AC or DC. For such a requirement, the type of switch and the contact materials are not particularly critical, and the only stipulation is that the switch should be adequately rated for the voltage and current required. The value of contact resistance, for example, will make only a small difference in the voltage across the closed contacts, and this difference will be negligible compared to the voltage that the switch is handling. The non-reactive load implies that there will no current surge on making connection, nor any voltage surge on breaking connection. Because of this, no special precautions have to be taken against arcing, and a long life can be expected. Only if special environmental conditions have to be allowed for will there be any difficulty in selecting a suitable switch from any supplier.

Rather more care needs to be taken if the switching has to cope with high currents at high voltage, and considerably more care is needed if surges exist at switch-on or at switch-off. Taking these one at a time, the handling of high currents requires large contact areas, and contact materials that are of low contact resistance. The dissipation of power at the contact is equal to I^2R, where I is the current (steady or surge) and R is the value of contact resistance. For example, for a current of 10 A and a contact resistance of 5 milliohms (mΩ) the power dissipated at the contact is $0.005 \times 100 = 0.5$ W. This may seem insignificant, but a lot depends on how large the contact area is, and how well the contacts can dissipate the resulting rise of temperature. Just to put these figures into perspective, the 5 inch, 1 kW hotplate on an electric cooker will have a total surface area of 12,667 mm^2, giving a dissipation of 0.07 W per mm^2. To achieve the same dissipation per unit area for switch contacts dissipating 0.5 W would require an area of contact of about 7 mm^2, corresponding to a good fit of contacts of diameter 3 mm. From all this, you can see that a contact resistance of only 5 mΩ for a 10 A switch is by no means negligibly low, and that switch contacts, if thermally isolated, would run very hot. Even if the voltage drop across the contact resistance is negligible, the effect of heat dissipation may not be, so that the size of the contact resistance and the area of the contacts are important considerations in any switch that handles large currents in the range of 1 A upwards. These factors will, of course, have been allowed for in the rating of the switch, but not all manufacturers will calculate their ratings making the same assumptions about thermal dissipation and the ratio of true contact area to measured area. If a large-current switch is intended for a sensitive application in which failure would entail considerable consequent damage (shutting down a complete installation, for example), then it would be wise either to use more generously rated switches, or to measure the temperature of a few samples after an eight-hour period at full rated current in the expected maximum ambient temperature. The contact resistance should also be measured before and after such a test, and any significant increase will point to local overheating, which will not necessarily show up in the temperature measurements.

Switching for low-voltage supplies brings another set of problems, depending on whether the current is large or small. Most low-voltage switching is likely to involve large currents, so that a low contact resistance is an important requirement both from the dissipation and the voltage-drop considerations. In addition, low-voltage DC supplies are likely to use large capacitors on each side of the switch, and the capacitor on the load side will charge at the instant of switch-on, causing a surge current that can be very large (at least ten times the average supply current). If this might cause difficulties, non-linear surge-suppressing resistors can be included in the supply line, but the voltage drop across such a device is usually prohibitive for low-voltage supplies. If possible, circuits should be arranged so that no large capacitors are connected on the load side of the switch between supply voltage lines. Unfortunately, the requirement for the minimum value of contact resistance in a switch for low-voltage, high-current applications often conflicts with the equally important requirement of long life. As always, there is a trade-off between contact resistance and contact durability.

The easiest requirement to fulfil for low-voltage supplies is for low current (in the range of a few mA to a few hundred mA). At these currents, contact resistance is not the main consideration, small contact areas can be used, and only the possibility of surge current or voltage can make for any difficulty in switch selection. The mechanical action of the switch is seldom important. There is no need for a snap action, so the full range of actions can be considered, and the physical size of the switch is unimportant from the electrical point of view. Switch size should, however be considered from the ergonomic point of view. Many very small switches are available, but they are decidedly difficult to use simply because they are so small. Long lines of such switches should be avoided, no matter how satisfying they may look, because it is highly likely that at some stage a user will unintentionally switch over two at once. When switch operating levers are so close that a human fingertip can cover more than one, trouble can be expected sooner or later.

Ratings and specifications

The published ratings for a switch will invariably start with the voltage and current rating of the contacts, showing both DC and AC ratings. These ratings should be taken as absolute maxima if the highest standard of reliability is required. For example, if a switch is rated at 30 V, 4 A DC and 250 V 2 A AC, then it is reasonable to use the switch with 30 V, 4 A either AC or DC, and up to 250 V 2 A AC, but not to exceed the absolute limits of 250 V AC or 30 V DC, and the absolute limits of 4 A at low voltage and 2 A at high voltage. You should not, in other words, work on the basis that you can trade voltage for current and assume that if the voltage is less than 30 V, then the current can be more than 4 A. As has been indicated earlier, the voltage rating and the current rating for a switch are imposed through different constraints, and a change in one quantity does not necessarily affect the other, unlike power dissipation in a load.

The other important quantity, which is not always quoted, is contact resistance. When this is quoted, it will be in units of milliohms, typically in the region of 5–20 mΩ. The figure that is quoted is usually the initial contact resistance, meaning the contact of the switch as supplied, since contact resistance tends to increase with the age of a switch when the switch is put into service. The rise in contact resistance will be greater if the switch does not use wiping contacts, if the environment is corrosive, and if arcing is frequent. Contact resistance is seldom of the greatest importance in non-signal switches, and for that reason is seldom quoted for these switches.

The reliability of a switch will be quoted in terms of the average number of operations to failure under accelerated test conditions. These test conditions may be unusually severe, so that for 'average' applications it is normal to find that the actual life of a switch is greater than the quoted amount. In many cases this may be quite irrelevant, because the equipment in which the switch is installed will have failed or been scrapped long before the switch fails. The expected life is usually quoted as two figures, the mechanical life and the electrical life. The mechanical life expresses the average number of operations before a mechanical failure, such as a broken spring or contact leaf. This is usually longer than the electrical life, which is the average number of operations up to the point of electrical failure through unacceptable contact resistance, contact welding, or failure to make contact. A typical figure for this quantity is 50,000 operations. To put this into perspective, if a piece of equipment is switched on and off once a day in a 5-day week cycle, the switch failure can be expected at about 96 years. In other words, a switch in a reasonable environment is not likely to be the component that determines the effective life of a piece of equipment.

The problem is that a switch, of all the components in an electronic circuit, is least likely to be located in a favourable environment. This applies particularly to front panel switches that will be exposed to shop-floor atmosphere and a good deal of heavy mechanical use. Whereas the other components are encased, possibly encapsulated, and do not, in the main, have moving contacts, a panel switch is located where the surrounding atmosphere can reach the contacts. It is also subject to large forces on its actuating levers. Manual operation of a switch can vary from gentle pressure to violent prodding of a push button or fast flicking of a toggle lever. The actuating mechanism must be strong enough to withstand the extremes of treatment that are likely, but must not, as far as possible, allow such treatment to affect the contacts. The provision of strong and freely moving actuating parts makes it difficult to seal the interior of a switch in any way, and because of the effects of arcing, sealing is seldom desirable in any case. Expected life is therefore a quantity that can be very misleading unless it has been measured under conditions that approximate to the working conditions of the switch.

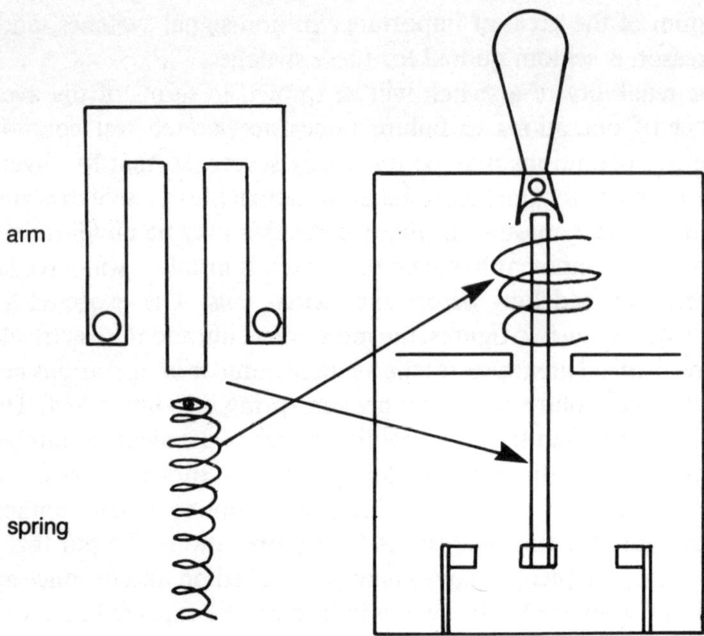

Fig. 2.2 The toggling action of a lever switch.

Snap-over action

The old-style knife switch had an action for which switching speed depended entirely on the operator, since the moving contact(s) would be mounted on the switch handle that the operator held. Modern switches for supply use do not rely entirely on the movement of the operator to determine opening or closing speeds, so that the switch consists mechanically of two parts, the contact part and the actuating part. Of these, the contact part consists of a spring-loaded toggle (Figure 2.2) which has two stable positions (the positions in which contact is made). In any other position of the toggle, apart from a precariously balanced central position, the spring action will ensure that the moving contact is in one stable position or the other. This spring also determines the force with which the contacts are held together.

The actuating portion of the switch will, in its motion, displace part of the spring toggle until the spring action flips over the switch. In this way the shape of the toggle and the strength of the spring determine the minimum switching time. No matter how slowly the switch is actuated, the toggle action should ensure that the moving contact snaps over, and no normal style of actuation should allow the moving contact to rest in any position except its contact (or on/off) states. The action of the operator can, however, affect the switching time in the sense that fast operation can speed up switching. Note that the action of a biased switch (see below) is very much more directly affected by the operator. The actuator of the switch does not affect the static contact pressure, but very fast actuation can greatly increase the *dynamic pressure*, the momentary pressure due to the momentum of the moving contact assembly. This can in turn lead to excessive *switch bounce*, in which the contacts close and then bounce open again before closing finally. Switch bounce is more of a hazard for signal-carrying switches, but it can cause increased severity of arcing in non-signal switches, and the excessive mechanical force that accompanies it can also shorten the life of a switch.

Contact configurations and actions

Switch contact configurations are primarily described in terms of the number of *poles* and number of *throws* or *ways*. A switch pole is a moving contact, and the throws or ways are the fixed contacts against which the moving pole can rest. The term 'throw' is usually reserved for mains switches, mostly single- or double-throw, and 'way' for signal-carrying switches. A single-pole, single-throw switch (SPST) will provide on/off action for a single line, and is also described as single-pole on/off. Such switches are seldom used for AC mains nowadays and are more likely to be encountered on DC supply lines. For AC use, safety requirements call for both live and neutral lines to be broken by a switch, so that double-pole single-throw (DPST) switches will be specified for this type of use. Double-throw switches are not so common for mains switches in electronics use – their domestic use is in two-way switching systems, and they have some limited applicability for this type of action in electronics circuits. A double-throw switch will also be described as a *changeover* type. The word 'throw', used to describe the number of fixed contacts, is a reminder of the snap-over action that is required of this type of switch.

Where a single or multiple contact is required to make connection to more than two fixed contacts per pole, then the fixed contacts are referred to as poles. This type of configuration is much more common in signal switches, and switches of up to twelve ways per pole are available in the traditional wafer form, as noted in Chapter 3.

The configuration of a double-throw switch also takes account of the relative timing of contact. The normal requirement is for one contact to break before the other contact is made, and this type of break-before-make action is standard. The alternative is make-before-break (MBB), in which the moving pole is momentarily in contact with both fixed contacts during the changeover period. Such an action is permissible only if the voltages at the fixed contacts are approximately equal, or the resistance levels are such that very little current can flow between the fixed contacts. Once again, this type of switching action is more likely to be applicable in signal-carrying circuits. There may be a choice of fast or slow contact make or break for some switches.

A switch may be *biased*, meaning that one position is stable, and the other, off or on, is attained only for as long as the operator maintains pressure on the switch actuator. These switches are used where a supply is required to be on only momentarily – often in association with the use of a hold-on relay – or can be interrupted momentarily. A biased switch can be of the off-on-off type, in which the stable condition is off, or the on-off-on type, in which the stable condition is on. The off-on-off type is by far the more common. The ordinary type of push-button switch is by its nature off-on-off biased, though this is usually described for such switches as *momentary action* (see, however, *alternate action* below), and biased switches are also available in toggle form.

Toggle switches

The toggle switch accounts for the majority of switch sales, and is available in the largest range of current and voltage ratings for electronics use. In addition to the large range of electrical characteristics, toggle switches are also available in a very wide variety of mechanical styles of mounting and actuation. The most common mounting method is the traditional threaded bush, generally of 0.468 inch diameter (11.9 mm) and 10–11 mm length. The bush uses two clamping nuts so that the body of the switch need not be placed hard against the panel on which it is mounted (Figure 2.3). Both clamping nuts should always be used, because it is bad practice to butt the switch

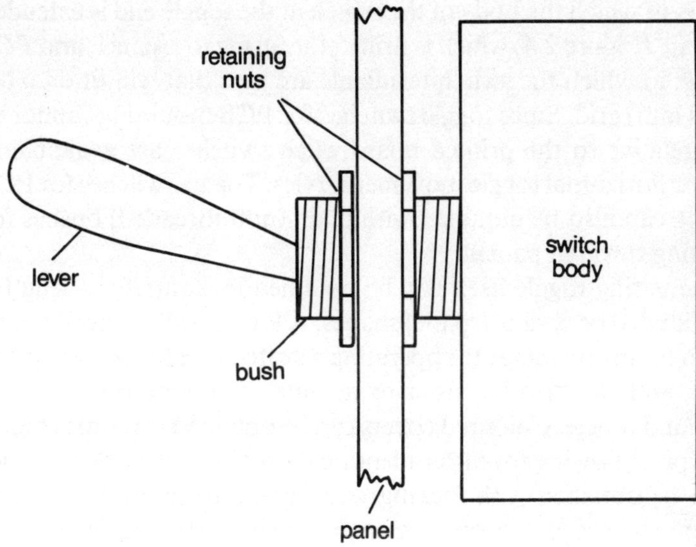

Fig. 2.3 The use of clamping nuts to ensure that switch fastening places no stress on the switch body.

body against a panel with a single retaining nut on the other side. The use of bush mounting allows the switch to be mounted in any position on a panel, but the conventional position allows the toggle to be operated in a vertical plane. Alternative mountings include *plate*

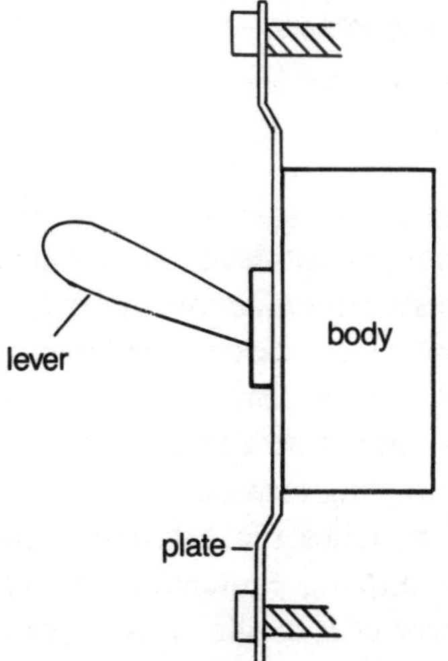

Fig. 2.4 Plate mounting of a switch – less common than bush mounting.

mounting, in which the body of the switch at the toggle end is extended into a plate (Figure 2.4) which is drilled for fitting to a panel, and *PCB mounting*, in which the switch terminals are pins that will fit on a 0.1 inch (2.5 mm) grid. Since toggle switches for PCB mounting cannot be rotated relative to the printed board, such switches are available in vertical or horizontal toggle movement styles. Toggle switches for PCB mounting can also be obtained with plain (non-threaded) bushes for push-fitting through panels.

The actuating toggle itself can be obtained in a variety of lengths, and in rounded or paddle (spade) shapes. A long paddle-shaped toggle can often be an advantage for operating a switch that has a high spring pressure, and can also be easier to operate in an emergency than a short, round toggle. Coloured covers can be obtained for some round-toggle types, allowing for faster identification of switches. Some types can also be obtained with locking lever action to prevent casual or accidental switching. To operate such a switch, the toggle must be pulled out slightly against the tension of a spring before the switch-over movement can be made. Waterproof sealing glands are available for some sizes, and PVC splash covers are also available for a lesser degree of protection. Connections are made by all the available methods of solder tags or buckets, screw fittings, PCB pins, and slide-on receptacles.

The contact materials for toggle switches will normally be silver-plated nickel, though the smaller switch sizes may allow a choice of solid silver or gold-plated contacts.

Rocker switches

Rocker switches are a more recent development in switch technology. The aim is to make the switching action easier from the ergonomic point of view: a rocker switch can be operated by a push action as well as by the conventional flicking action, or by any combination of these movements. The rocking actuator of the switch is much more closely coupled to the internal toggling arms, and will snap over at the same rate. For small rocker switches the external rocker is part of the assembly that bears the moving contact(s). The actuating rocker is much wider than the lever of a toggle switch, even a paddle-shaped lever, and results in lower pressure on the operating finger. This extra

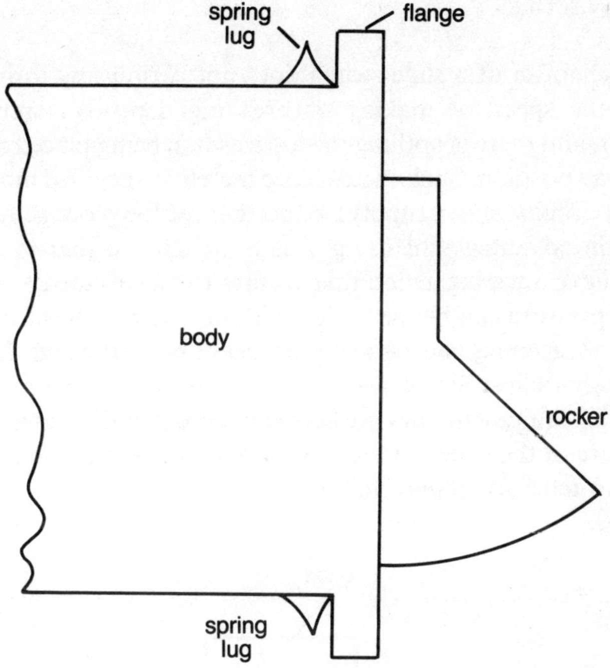

Fig. 2.5 The use of spring lugs to retain a switch in a rectangular hole of precise dimensions.

width requires a broader switch body, so that rocker switches often allow space for signal lights. The considerable difference in the shape of the actuator means that bush mounting is not used, and rocker switches are for the most part fixed by lugs that have a snap action against the mounting panels (Figure 2.5), though a few types use mounting plates that can be bolted into place.

The mechanism of some types of rocker switch can make it possible for the operator to hold the switch in a 'neutral' position. If this would be undesirable such switches should not be used, though this does not preclude the use of other designs of rocker switch. The range of electrical and mechanical parameters is much less than for toggle switches, but rocker switches do have particular advantages: they are available in colours to match panel schemes, and can be obtained illuminated. Illumination in non-signal switches generally makes use of a miniature neon with a series resistor of a size suitable for 240 V AC operation (and not suitable for low voltage use).

Slide switches

The mechanism of a slide switch does not permit any form of snap action: the speed of making or breaking depends purely on the operator, and there is nothing to stop a switch being placed and left in a mid-way position. Such switches are therefore specified more for use in signal circuits, and in supply circuits that use low-voltage AC or DC. The main advantages of using a slide switch are that the contacts usually have a wiping action that ensures clean surfaces, and that the contact pressure can be quite high without making the switch action difficult. Fastening can be achieved using bolts through lugs, PCB mounting or clips. Single slide switches are becoming a rare sight in electronics equipment; they are being replaced by DIL pattern switches (which are of the slide pattern) for signal voltages, and by miniature rocker switches for supply voltage use.

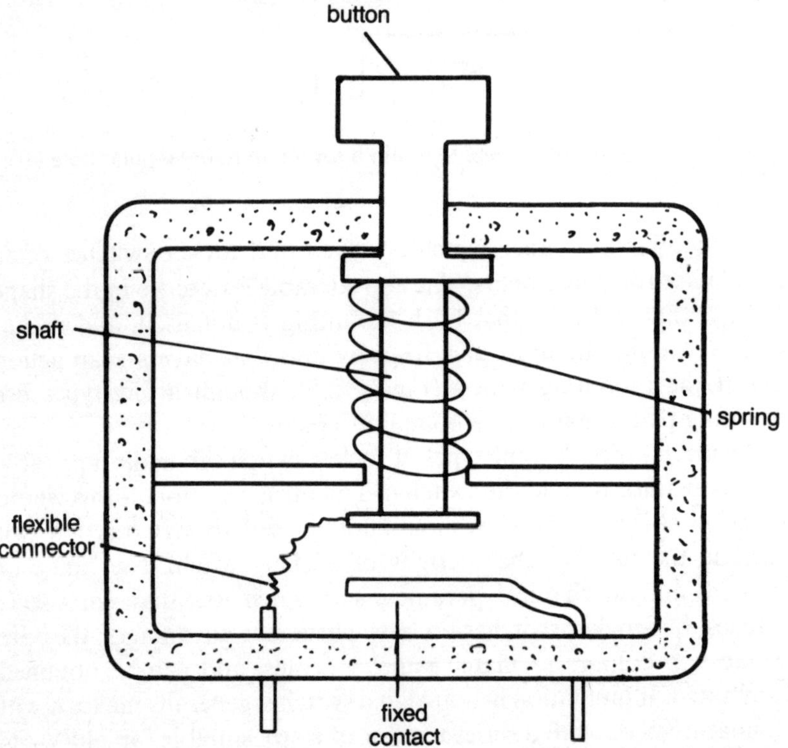

Fig. 2.6 Operating principle of the momentary-contact push-for-on switch.

Push-button switches

Push-button switches challenge the range of toggle switches for variety, though many push-button types are intended purely for signal uses. The two basic types of push-button switch are the *momentary-action* and *alternate-action* types. The internal construction of a momentary-action switch (Figure 2.6) demonstrates the principles. The moving contact is held on the shaft that is actuated by the button, and the contact is made (in this example) for as long as the button is pushed. Releasing the button allows the main spring to separate the contacts, subject to the range of travel of any internal spring that is used to buffer the moving contact from the shaft. This type of momentary-use switch can be of push-to-make or push-to-break varieties.

The other basic type is the alternate-action type. In this version, pushing the button will reverse the effect of the previous push, switching either on or off. The mechanism is usually rotary and permits only fairly small contact pressures, so these switches are not intended for high current uses. For the mains switching versions, 250 V 3 A is a typical rating, though higher current ratings are available. The term 'latching' is often used of the alternate mechanism, but latching has a specialised meaning when applied to a push-button switch. A latching push-button switch will allow the button position to indicate whether the switch is on or off, for example by having the button fully out when the switch is off, and partly in when the switch is on. The mechanism of a latching switch is basically that of the alternate type, but with the push button connected to the internal mechanism rather than simply pushing.

Push-button switches of either basic type are available with solder lugs or pins, or PCB mountings, but seldom with snap-on receptacles. Like rocker switches, many types of push-button switches are available with illuminated buttons in a variety of shapes and colours. The illumination is usually by miniature filament bulbs that will require a low-voltage supply, rather than by the neon which is used for rocker switch types.

Rotary and key action

The most common type of rotary switch is the wafer type (this is dealt with in the following chapter along with other signal circuit switches), but for more specialised purposes rotary mains switches are available. A rotary action switch is often found incorporated into a potentiometer, providing the combined on/off/volume control action that was very common at one time for mains radios and TV receivers, and is still found on some audio equipment. Separate rotary action mains switches are still available, but mainly for replacement purposes. One modern type is intended to discourage casual use by having a shaft with a screwdriver slot, but the main use for rotary mains switches in supply circuits nowadays is to permit key action for security purposes. The variations in key mechanism include removable or trapped keys, and common or random key patterns. The use of a trapped key allows a circuit to be switched on by a keyholder, but since the key must then remain in the switch the equipment can be switched off by anyone, as safety regulations will usually stipulate. If the key can be removed, some other switch-off mechanism in the form of an emergency button should be provided.

Microswitches

The microswitch is a component with a comparatively short history but a rapid development in use. The 'micro' part of the title is slightly misleading – these switches need not be particularly small physically (though some are) and often have contact areas as large as conventional toggle switches. The distinguishing feature of a microswitch is its toggling action, which calls for a very small amount of movement of the actuating button or lever. For a basic microswitch operated directly the movement will typically be 0.4 mm before contacts are changed over, with a *differential* of 0.05 mm. The differential is the minimum change in operating-pin position between on and off, and it is this quantity in particular that is responsible for the name of the switch. The mechanism also permits the operating pin to move beyond the position required for switching over, typically for 0.15–1.5 mm depending on type. These small actuating movements are achieved along with voltage and current ratings of mains standard – typically 480 V, 15 A.

The microswitch itself is operated by pushing a pin of small diameter (typically 1.5 mm), so environmental sealing is much easier than it is with other types. In addition, a variety of external mechanisms can be used to actuate the pin. The operating forces call for quite high

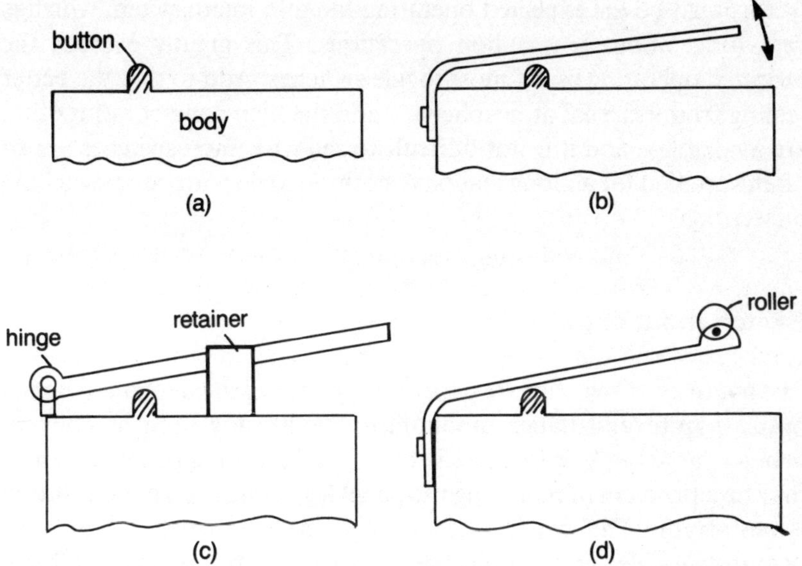

Fig. 2.7 Microswitch actuation (a) plain button, (b) leaf, (c) lever, (d) roller leaf.

pressures on the pin of a conventional microswitch, so that the external addition of buttons, plain levers, and roller-arm levers (Figure 2.7) all make operation easier. The plain pin type is used for door interlock switching, and in any other application in which other moving parts can apply pressure directly. By using push buttons, operation by finger pressure is made easier; by using levers, lighter action at the expense of larger amounts of movement can be achieved.

Types of low-torque microswitch can be obtained, allowing mains switching at up to 5 A but with such low torque operation that the action can be initiated by a small vane blown in an airstream, the weight of a coin, and so on. For use with signals or with low voltage supplies, subminiature microswitches can be obtained that combine very small size with the usual short operating travel, and in this case with small amounts of force as well.

The main applications for microswitches when they originally appeared were as door safety switches and alarm systems. The action is always of the momentary variety, so that microswitches are employed as main unit switches only in conjunction with hold-on relays, since there is no latching action in the microswitch. This has not deterred the rapid spread in the use of microswitches, some with mechanical latching mechanisms attached externally. One important factor must be the expected operating life of a microswitch, which is very long: about ten million operations. This greatly exceeds the expected operating life of most toggle switches. Add to that the better sealing from external atmospheres – and the high contact ratings that are available – and it is not difficult to see why microswitches are so often specified for actions that have nothing to do with door switching or security.

Switch mounting

The majority of switches are panel-mounted, and the methods (such as bush, snap-fit and flange mounting) have already been mentioned. When a large bank of switches is to be available on a panel, but there may be a problem of removing the panel for servicing, a good solution is a sub-frame. The switches are mounted on a sub-panel, with only the operating toggles or other mechanical parts protruding through holes in the main panel, and the sub-panel is attached in only a few places. In this way, only a few bolts need to be removed to allow the main panel to be detached, avoiding the need to remove each switch individually. A few types of switch require specialised mountings – the rotary edge-wheel type, for instance, is very difficult to accommodate. Microswitches in particular do not lend themselves well to panel mounting, and usually require the use of angle brackets. In general, any application involving the use of microswitches will need some thought as to fixing methods, since the built-in fixings consist only of two holes in the casing of the microswitch that do not permit direct connection to any surface not parallel to the movement of the actuating pin.

Chapter Three
Signal-carrying Switches

Special requirements

A switch intended to carry signals, as distinct from DC or AC supplies, must inevitably be subject to several requirements in addition to the normal needs already described. The type and frequency range of the signals that the switch is required to carry determine to a very great extent the special needs that will prevail, because a signal-carrying switch for low-frequency AC in the range 5–240 V and with currents that are not less than 50 mA or so will be identical, to all intents and purposes, with a switch for AC supply use. The important differences in design appear when a switch is intended to carry signals of high frequency, or signals in which rapid changes of voltage are important, like digital signals. A feature of signal switching that is of no importance in supply switching is that the switch can itself add noise to the signal or even generate spurious signals. Capacitance between contacts (and from contacts to earth) is of considerable importance, and the maintenance of low contact resistance becomes more difficult just as it becomes more important.

Contact resistance can be particularly important if signals of low voltage level are being switched. At low signal voltage levels there is no sparking or arcing which might expose fresh surfaces now and again, and unless a wiping action can be obtained, as is often the case, the contact resistance can become quite high. This in itself is not necessarily a hazard if the signal current levels are low, but if the switch is used for low voltage levels at which current levels are not low, then the voltage drop across the switch contacts will be appreciable. Since the greatest problem in low-level circuits is usually the maintenance of an adequate signal-to-noise ratio, any impedance in series with the signal current is very undesirable. For low-level circuit switching, gold-plated contacts are usually recommended. Though the resistance is never as low as can be obtained with silver, gold has the great advantage of being chemically inert, so that contact resistance is very stable. The softness of the metal also tends to ensure a greater area of contact, which compensates for the higher resistivity.

Other important contact considerations are capacitance and non-linearity. If the contact resistance is high, then a thin non-conducting film probably exists on the contacts, and this film will form a capacitor coupling between the contacts. For signals at very high frequencies, the reactance of this capacitive coupling may be lower than the resistance of the contacts, but like any other reactance will introduce a phase shift. Because of the low reactance of the layer, an increase in contact resistance may be unnoticeable in terms of signal strength, but the phase changes can cause problems which will not necessarily be attributed to the switch. For digital signals at lower frequencies which have short rise and fall times, the capacitive coupling will cause some differentiation of the signals, which may have no noticeable effect until it causes loss of amplitude.

A rectifying contact is considerably more troublesome. This can arise where there is metal to oxide contact on a switch, and in some cases can be the result of severe arcing which has left a sharp whisker of metal making contact against an oxidized surface on the opposite contact. The result is a diode whose forward resistance and back resistance values are not very greatly different, but which differ nonetheless. If appreciable signal currents flow across this part-rectifying contact then the diode action will have the usual effects on the signal. One effect will be distortion of the signal, such as differences between positive-going and negative-going portions of the signal. The other effect, which can be very serious, is DC bias. If the switch feeds a DC-sensitive input, such as a directly-coupled input to an operational amplifier, the amount of DC bias produced by the rectifying action of the switch contacts may be enough to bias the amplifier off or into non-linear action. For digital circuits, the effect can be to prevent any signal input from being effective. Modern contact materials make this condition unlikely, but there is always a possibility of a rectifying action arising through contamination, particularly in humid atmospheres, and its effects are seldom immediately blamed on the switch contacts.

For the rather special case of RF switching at high power, the problem of arcing is much more severe than it is for supply current switches. This is because the time of opening the switch contacts will inevitably be greater than the time of a cycle, so that the waveform will be at its peak voltage at least once while the contacts are opening. Since the contacts have capacitive reactance in any case, a considerable current can flow, and this will cause ionization of the air which, because of the short time between peaks of a high-frequency signal, cannot easily be damped out. Switches for high-power, high-frequency applications are therefore a very specialised branch of switching, and one for which a one-off switch may have to be designed to suit the particular application.

Speed of action

A mechanical switch operating with mains AC will normally have an opening or closing time that is short compared with the time of a cycle. This does not apply to switches that are used for higher frequencies, so that the rating of switch contacts for such signals is closer to the DC rating than to the AC (mains) rating. If switching has to be carried out in very short times, then semiconductor switching devices need to be used. These are dealt with later in this chapter, though more as adjuncts to mechanical switches than as a subject in their own right. Since switching speed is inevitably slow in comparison to the time of one cycle of wave, it is less relevant to the specification of switches for signal frequencies, and for low-power signals the time of switching is usually dependent on the action of the operator. A wafer switch, for example, will change connections in a time that is determined completely by how fast the operator turns the switch knob. For audio work, switches with a snap-over action are often used, but for all sorts of work where conditions are critical in terms of speed and noise-free operation the combination of a mechanical switch with a semiconductor switch is now more common.

Capacitance

Two figures for capacitance are significant in a switch intended for use with signals: the stray capacitance to earth and the capacitance across the open contacts. Taking the latter first, the open-circuit capacitance (Figure 3.1) determines to a large extent the efficacy of the switch in

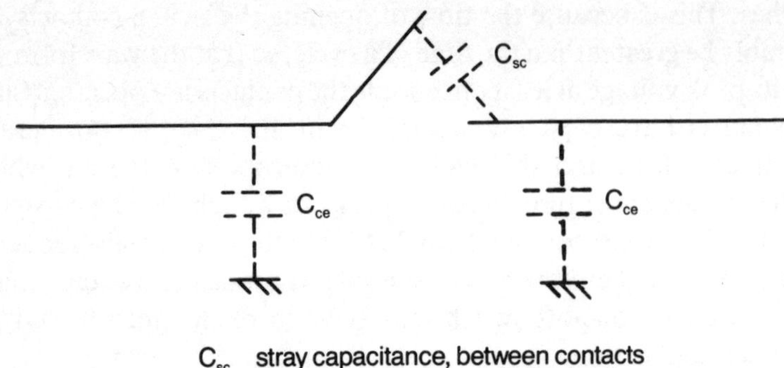

Fig. 3.1 The stray capacitances that exist across switch contacts and from switch contacts to earth.

terms of open and closed circuit impedance. The figure, unfortunately, is seldom quoted but will be in the range of 10–20 pF for the common types of signal switches. This will cause a significant amount of signal to appear on 'open' contacts, particularly for high-frequency or fast-rising signals, unless the switches are operated in circuits of low

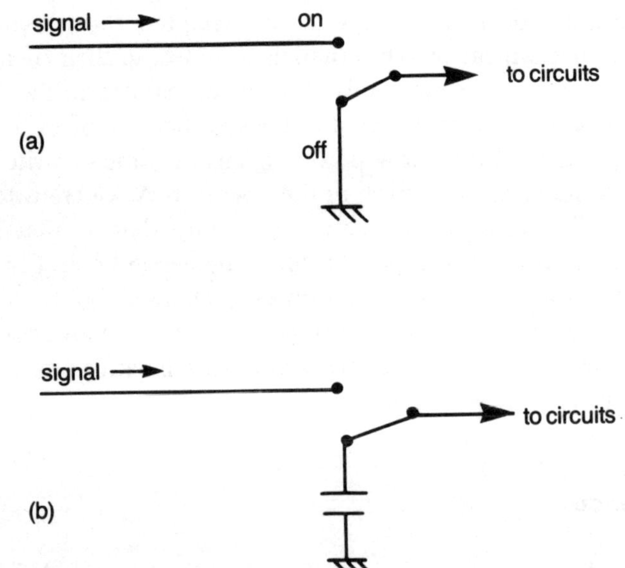

Fig. 3.2 Earthing the 'cold' side of a switch to reduce the effect of signal coupling through stray capacitance. (a) No DC level present at switch, (b) use a capacitor when DC is present.

impedance. For this reason, a simple open and close type of switching arrangement is seldom suitable for high-frequency signal switching, and the switch should ensure that no unwanted signal is coupled across by earthing the 'cold' side of the switch when the switch is off. Obviously if a DC bias exists on the cold side of the switch contacts the earthing will be done through a capacitor. Figure 3.2 illustrates this principle applied to a single-way switch.

The type of switch mechanism very greatly affects the amount of capacitance that exists between the contacts. The contacts of a wafer switch lie in the same plane, with only a very tiny area of contact conductor parallel to any other conductor, so they have very low capacitance across the contacts. This is a major reason for the continued popularity of this switch type, which has remained almost unchanged (apart from improvements in materials) for some sixty years. The type of switch in which open contacts lie parallel to each other is the worst from the point of view of open-contact capacitance, though if the open contacts are reasonably far apart, the amount of capacitance between them can be as low as can be achieved in a wafer design. Since speed of action is seldom an important factor for the mechanical switch, a comparatively large distance between open contacts can be specified in push-button and keyboard switches.

Another important aspect of capacitance in signal-carrying switches is the stray capacitance between contacts that carry different signals. If a switch is used to carry several signal channels, then capacitance between contacts can lead to signal crosstalk. If impedance levels and signal levels are both low, the amount of such crosstalk may be insignificant, but in the worst-case example of a high-level signal on one channel and a low-level signal on another, with comparatively high impedance, the amount of crosstalk would become intolerable. Multi-channel switching of this type is best handled by ganged assemblies in which the channels can be kept physically well separated. Once again, the familiar wafer switch copes best with this requirement if one wafer is used for each channel, with earthed screens placed between the wafers.

Stray capacitance to earth is, by contrast, much less of a hazard unless the switch is being used in a high-impedance circuit. By their nature, most switches are mounted on conducting panels, so that stray capacitance to earth is always significant, but since this stray capacitance has little or no effect on switching performance it can be treated like any other stray capacitance to earth in the circuit. Stray capacitance of this type becomes a problem only if the local earth happens to be the hand of the operator, which can occur when a signal-carrying switch is mounted on a non-conducting panel. This will have the effect of making the stray capacitance to earth much larger when the switch is being operated than at other times, and is the same effect as the 'hand-capacity' problem that haunted amateur radio builders many years ago. Once again, the problem may be negligible: it is only likely to cause difficulties if the switch happens to be part of a tuned circuit whose frequency can be changed by the alteration in capacitance to earth. This problem can also be solved by the use of a wafer switch with an earthing screen. In general, switches should not be mounted on insulating panels for any application in which variation of capacitance is undesirable.

All these remarks on stray capacitance underline the importance of good circuit design in which switches are never placed in high-impedance circuits. Except for the most demanding of waveforms, with very fast rise and fall times, switch capacitance is never a problem when impedance levels are low. At high impedance levels, the switch capacitance to earth is just one of the stray capacitances to earth, and presents no special problems unless the capacitance varies. The main problem for higher impedance circuits is capacitance between switch contacts, and this can be reduced to very small amounts, as we have seen. The usual and least avoidable type of high-impedance circuit is the tuned circuit in which a switch is used to change inductors, capacitors or the complete LC tuned circuit for an oscillator. In this case, the stray capacitance across contacts has an effect that is constant, and which can be dealt with by any form of fine-tuning adjustment.

Noise

All signal noise is caused by random motion of electrons or other current-carriers in a circuit. Even in a piece of wire consisting of the same metal throughout there will be some noise caused by the irregular movement of the electrons at room temperature. This type of noise can be greatly reduced by lowering the temperature, but is usually negligible compared to the other noisc sources in a circuit. In general, noise is generated mainly when a small number of electrons are moving comparatively rapidly. The main sources of noisc are therefore semiconductor junctions and bad contacts, and it is this latter source that is the main concern for the switch user. In addition, however, noise can be generated at junctions that are not semiconductor junctions. Wherever any two different metals meet there will be a build-up of electrons that causes a measurable voltage difference, called the *contact potential*. This contact potential is mainly DC, but because of irregularities in the movement of the electrons around the contacts it will have an AC component, which is noise.

Any type of switching circuit will therefore introduce noise into a signal that is switched. The noise arises to some extent from variations in contact resistance, but can also be caused by the fluctuating contact potentials if the contacts are not clean and of identical chemical composition. A switch is not a major noise-producing component in circuits unless it has been used in a circuit that operates at very low signal levels. Switching at such low levels is unusual, because it is normally possible to amplify a signal before switching it, but if such switching is unavoidable, it should make use of switches with gold-plated contacts. In addition, if switching at low signal levels is completely unavoidable, no DC should pass through the switch, because DC flow is by far the most likely cause of noise in a switch (as it is in a potentiometer).

The type of noise that is caused by the switch contacts while they are closed is the usual broad-band noise, but there can be an additional problem of impulse noise when switch contacts close, caused by the charge and discharge of stray capacitance by the DC levels in the circuit. A familiar effect of this at one time was the noise pulse that accompanied audio circuit switching, for example when changing an amplifier input from disc to tape. The root cause of this problem is the presence of DC on one of the switch contacts, or the use of different DC levels between contacts. The easy way out of this is, of course, to design the circuit so that switching is carried out at zero DC level at all times, but this may require the use of more coupling capacitors in the circuit path than is thought desirable, and in any case if the switch contacts are isolated to earth there will be a build-up of DC through the leakage resistance of any capacitor. The usual cure, if a purely mechanical switch is to be used, is to equalise the DC levels at the switch contacts, usually by the use of high-value resistors (Figure 3.3). This will lead to some signal breakthrough unless the signal circuits are at low impedance.

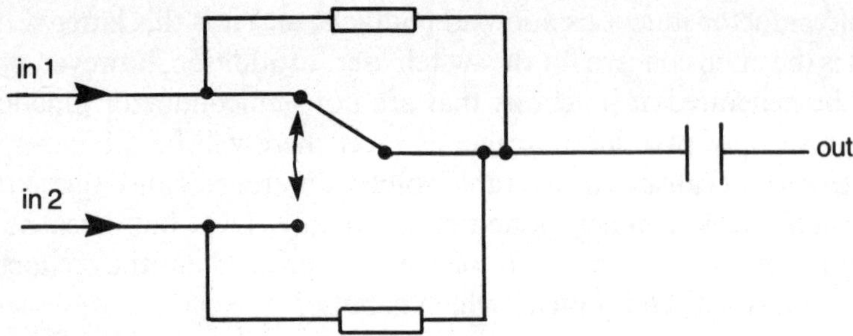

Fig. 3.3 Using leakage resistors to equalise DC levels when a switch is connected to a capacitive circuit.

Wafer switches

The familiar wafer switch still fulfils most of the requirements for signal switching, particularly in low-power circuits. The layout of a wafer (Figure 3.4) ensures minimum capacitance between contacts, and also means that the contacts are cleaned by the wiping action each time the switch is used. Because of the wiping action, silver-plated contacts can be used to provide low contact resistance (and low noise) with less risk of contamination than arises when the contacts are non-wiping. The main virtue of the switch, however, is that it can be constructed to order, for example allowing wafers to be widely spaced and separated by earthed screens if this is desirable. This is possible because the rotation of the inner part of a wafer (the *rotor*) is controlled by a shaft that is located by a flat side, and this shaft can be cut to whatever length is needed or joined to other shafts.

The mechanical control of the switching shaft is carried out in the section that is mounted on the panel. The switching positions are most usually at 30-degree intervals, and the movement of the shaft is regulated by a detent mechanism (Figure 3.5) which helps to ensure that the rotating contacts are not left between fixed contacts. The detent can also assist in speeding switching times, though this is not an important factor. Careless use of a wafer switch can leave contacts isolated between detent positions, but modern mechanisms are surprisingly efficient – surprisingly, that is, because the mechanisms are comparatively simple.

The main points that distinguish the wafer switches of today from those of some sixty years ago are construction and materials. The old type of wafer switch came as a ready-made item, using wafers of

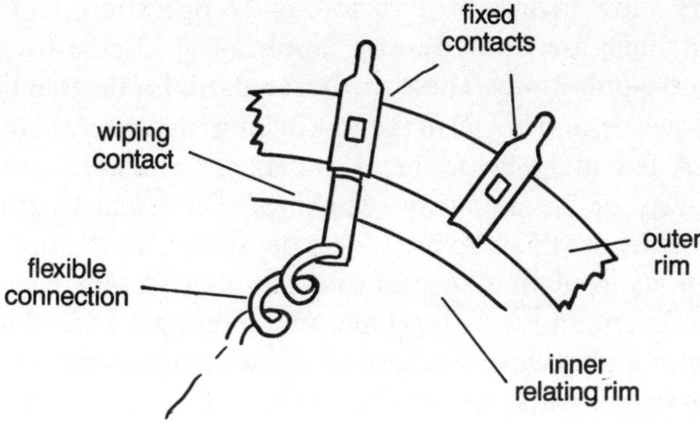

Fig. 3.4 The basic construction of a wafer switch. The use of contacts in one plane ensures that stray capacitances are very low, and the construction also ensures high resistance values.

laminate plastic (usually of the Bakelite family). The changes in wafer switches over the last twenty years have included the provision of switch kits that allow the user to specify the type and positioning of wafers, and of materials such as diallyl phthlate, usually glass-fibre filled, and acetals for exceptionally high insulation resistance coupled with considerable mechanical strength. The traditional form of wafer switch is still used to a very great extent, but modified wafer shapes for PCB mounting are now available. These, however, inevitably have higher capacitances between contacts than the traditional type of wafer because of the length and close spacing of the tracks to the contact pins.

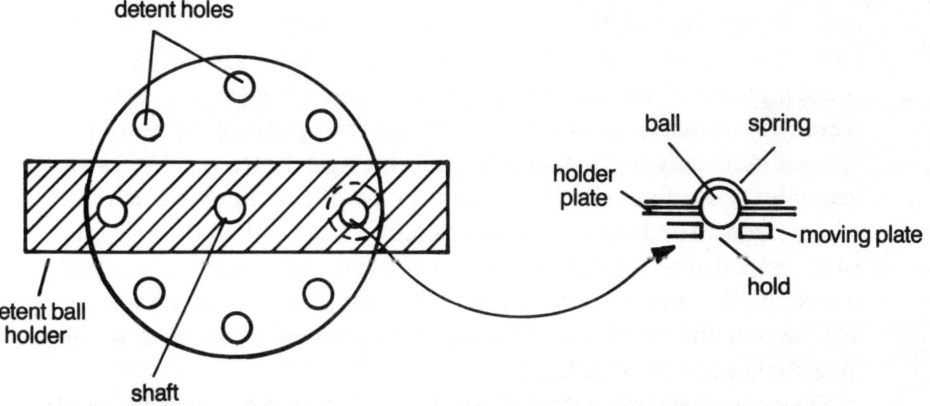

Fig. 3.5 The principle of the spring-loaded ball detent mechanism for wafer switches.

Wafers allow mainly break-before-make operation, and the standard switchings are 1-pole 12-way, 2-pole 6-way, 3-pole 4-way, 4-pole 3-way, and 6-pole 2-way. These are the standards for the traditional type of wafer switch, and not all of these switchings may be available in PCB wafers. A few make-before-break wafers are available, usually as 1-pole 11-way or 2-pole 5-way switchings. The usual length of shaft allows for the use of six to ten wafers depending on the size and type. In addition, insulating spacers and earthing screens can be fitted between wafers, and some types allow for a dummy wafer that consists of a stator only, with contacts to allow components to be wired between wafers without recourse to the main PCB. Another useful feature of most of these wafer kits is that a miniature mains switch can be fitted at the end of a shaft to allow for mains switching in addition to signal switching. This is not desirable if the wafers handle low-level or high-impedance signals. At the switch mechanism end, a key lock can be fitted to allow the switch to be locked in positions at 60-degree angles – note that this allows only alternate contacts to be used, since the conventional wafer contact positions are at 30 degrees.

Reed switches

Reed switches are widely used for signal switching, particularly at the lower signal frequencies. The main attraction of the reed type of switch is that the contacts are enclosed, and cannot therefore be damaged in adverse environments. In addition, the switch is magnetically operated either by a solenoid coil or by a permanent magnet, so that remote control is possible and variable capacitance to the hand of an operator is not a problem. In this respect, the reed switch is normally used as a secondary switch or relay in the sense that the operating current to its coil will normally be controlled by another switch. Reed switches are generally classed as relays for the purposes of cataloguing electronic components. The reed switch can therefore be mounted directly on a PCB deep inside the circuit, possibly in an inaccessible place, and operated from any conventional type of low-voltage switch on a panel. This avoids the problems of bringing a signal cable out to a panel-mounted switch and back again.

The basic reed principle is illustrated in Figure 3.6. The thin metal reeds are sealed into a glass tube, and bent so as to make contact because of the effect of a magnetic field. The usual arrangements of

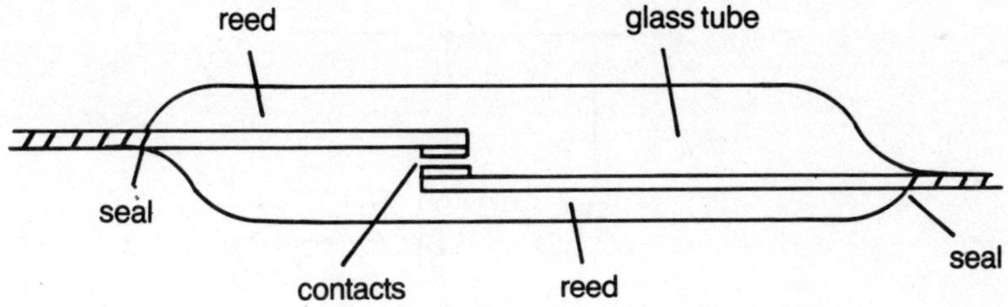

Fig. 3.6 A reed switch (or relay). The reed arms are usually made from a nickel alloy that will seal into glass.

contacts are normally open or changeover, and the reed portion can often be specified separate from the actuating coil or magnet. The use of the solenoid coil allows a relay type of operation, but the use of a permanent magnet allows mechanical operation (by moving the magnet to or from the reed tube) which can be custom-made to whatever pattern is needed. The typical magnet distances are of the order of 7–11 mm to operate the reed, 13–16 mm to release, so that a movement of around 5–20 mm would normally be used to ensure reliable opening and closing of the reed circuit. The predominant use of reeds, however, is with the solenoid, using up to 100 ampere-turns to operate the reed.

One important feature of reed-switch use is the fast switching time. This will be of the order of 1–2 ms to make, and can be as low as 0.2 ms to break, with reed resonant frequencies in the order of 800–2500 Hz. Fast switching speed is not necessarily an important feature of signal switches, but if signals have to be sampled, or if there has to be switching from one signal source to another at frequent intervals, then the reed type of switch has considerable applications. An important further consideration for high-frequency or fast rise-time signals is the very low capacitance between open contacts, which can be less than 1 pF for the normally-open type of reed. The use of silver or gold-plated contacts allows low contact resistance values of around 100 mΩ, and for applications that demand very low and stable contact resistance, mercury-wetted reeds can be obtained. The mercury-wetted type also has the advantage of providing very low-bounce operation, a point that will be taken up later.

The other features of reed switches are high insulation resistance of the order of 100,000 M (because of the use of glass encapsulation), and high breakdown voltage of at least 200 V. The operating currents are also quite surprisingly high for a device with small contacts and limited

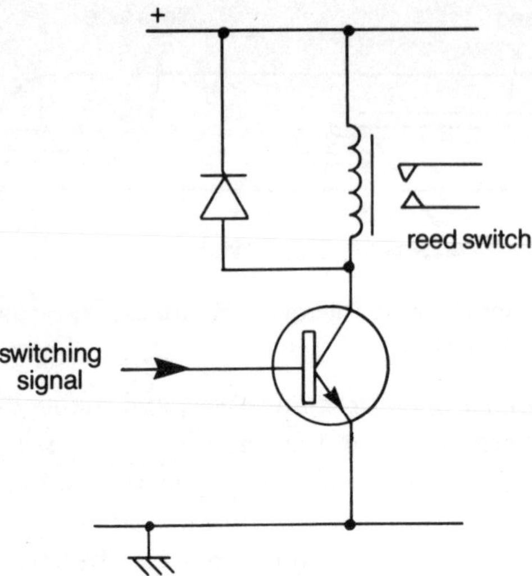

Fig. 3.7 Using a diode to protect a transistor from the effect of a reed-solenoid load.

contact force, of around 0.25–2 A. Mechanical life, depending on type of use and reed construction, ranges between one million and one hundred million operations.

The normal operation of the reed relay by its solenoid coil can raise one problem which is common to any device that uses an inductor. When the current through the coil is broken, the usual back-EMF is generated, and this can cause sparking between switch contacts (of the switch used to operate the solenoid) or breakdown of a transistor if this is used as the coil driver. When a reed switch and solenoid are constructed as one unit, the coil will usually incorporate a protection diode to avoid the effects of the back-EMF, and this implies that the polarity of connections to the coil will have to be observed. This in turn means that such a coil should not be used if you are operating the coil from an AC supply (as a 100 Hz sampling circuit, for example). If you are using a separate coil and reed, you will need to incorporate the diode for yourself (Figure 3.7) if you are using a DC supply to the coil.

Solid-state switches

The subject of solid-state switching could fill several books, and in this chapter we are concerned more with selection and use than with

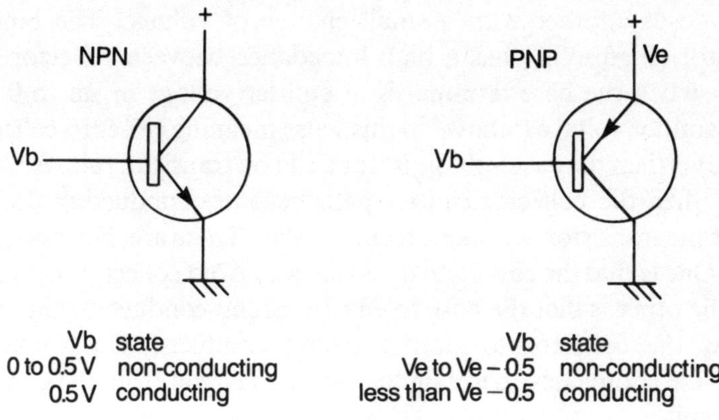

Fig. 3.8 Bipolar transistors used as switches. The disadvantage of using a bipolar transistor is that the operating input (to the base) is not isolated from the switching terminals (emitter and collector) and requires some current.

theory. In addition, the more specialised types of electronic switches will be dealt with in the following chapter. The fundamental principles of solid-state switching are that a transistor (bipolar or FET) has a low-impedance state and a high-impedance state, and can be switched from

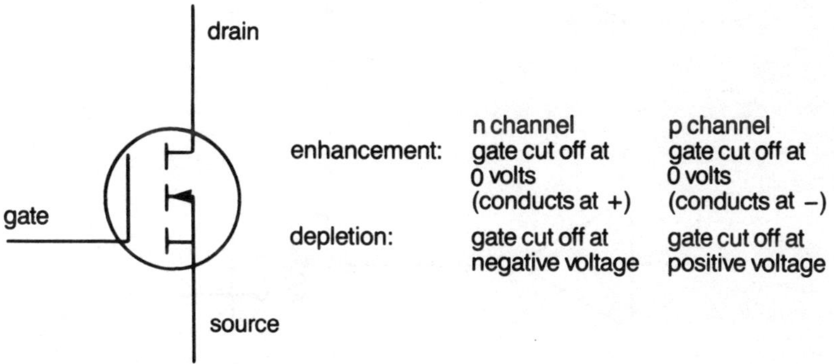

Fig. 3.9 The N-channel FET used as a switch. No current has to be supplied to the operating terminal (gate), and isolation of input from output can be very high, particularly in a bridge arrangement.

one state to another with a small change of voltage. The bipolar transistor (Figure 3.8) has a high impedance between collector and emitter when the base terminal is at emitter voltage or up to 0.5 V above emitter voltage ('above' in this sense meaning closer to collector voltage). When the base voltage is (for a silicon transistor) above about 0.55 V then the collector-emitter path becomes conducting. In this respect the transistor acts like a form of relay. There are, however, two snags. One is that the conductivity is one way from collector to emitter only; the other is that the base to emitter circuit conducts at the same time as the collector to emitter circuit conducts. This makes it impossible completely to isolate the switched circuit from the controlling circuit.

The use of bipolar transistors in signal-switching circuits is therefore a matter of careful design, making use of bridge circuits. Since the base voltage can be DC and can be decoupled, the bipolar transistor bridge can be used to switch signal circuits by means of a low voltage DC supply to the bases. In general, though, bipolar transistors are used more for DC switching than for signal purposes. This is because the

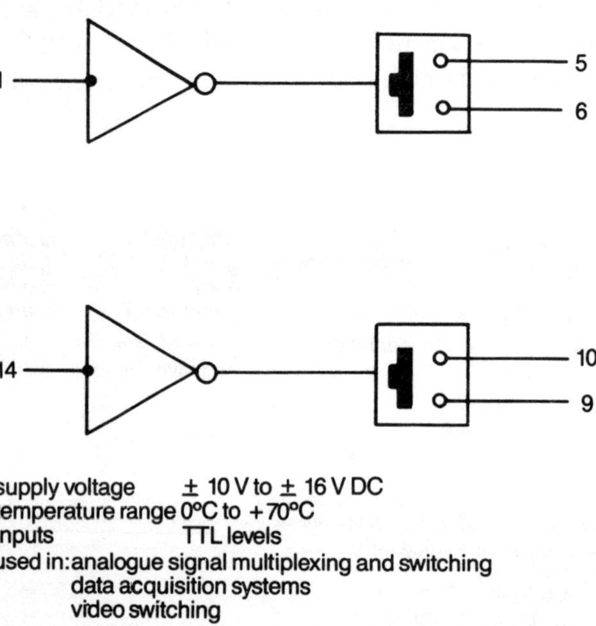

Fig. 3.10 The HI200 FET switch IC, which features low 'on' resistance that is almost constant throughout the frequency range, and high 'off' resistance. Switch transients are very small.

minimum collector to emitter voltage is around 0.2 V, so that signal amplitudes of less than that amount do not pass through the switch. For low-level signals, then, this causes an unacceptable amount of signal waveform distortion.

The use of FETs, particularly the MOSFET type, for signal switching is much simpler and more frequently encountered. The resistance between the source and the drain (Figure 3.9) is governed by the voltage at the gate, and the gate is almost completely isolated electrically from the other two electrodes. In addition, current will flow between source and drain even when the voltage difference is very low, and the geometry of the device can be such that current will continue to flow even if the polarity is reversed (provided that the gate voltage is maintained). Once again, bridge or two-transistor circuits are normally used, and the complete arrangement is available in IC form, usually as dual or quad single-pole single-throw switches. Figure 3.10 shows some typical characteristics, courtesy of RS Components. Switches of this type can be used with fast-changing waveforms such as video signals, and have only negligible switching transients. They are also widely used in audio preamplifiers because they permit switching that is free of the usual transients, and are extremely useful for switching data signals.

Chopper switches

At one time, vibrating mechanical switches were used for DC to AC conversion, particularly for car radios in which their use posed a

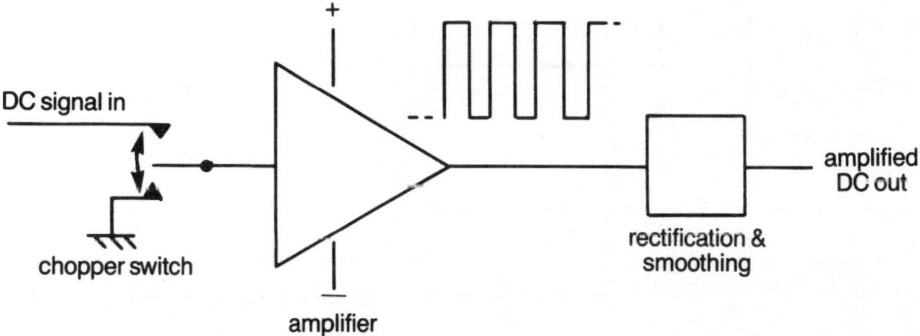

Fig. 3.11 Principle of a chopper amplifier for amplification of DC and very low-frequency signals.

filtering problem that was never completely overcome. With the coming of transistors to permit low-voltage operation of radios, the main applications for the technique of chopping were in DC amplifiers, using the principle illustrated in Figure 3.11. The input of an amplifier is connected alternately to the small DC signal and to earth, so that the input to the amplifier is a square wave at the frequency of operation of the switch. The early mechanical vibrating switches were often used synchronously (Figure 3.12) so that no rectification was needed, but more modern chopper amplifiers have used diode bridges as rectifying units. The important point, however, is the type of switching used. Reed switches were a useful intermediate stage, but only the use of modern FET amplifiers permitted the use of reasonably high frequencies for chopper amplifiers. An offshoot of this type of circuit has been the 'DC-DC' transformer, which consists of a DC to AC chopper, amplifier and rectifier in one single package. Such circuits are used extensively in digital circuitry where the main supply is +5 V but a few units need a low-current 12 V supply, or possibly a negative supply.

Digital switching

The switching of digital signals is a more specialised topic than analogue switching because of the range and nature of digital signals. Excluding the switching of supply voltages, switching actions for digital circuits can be of three separate types. First of all, a switch may be used to provide a signal into the circuit. An example of this is a push

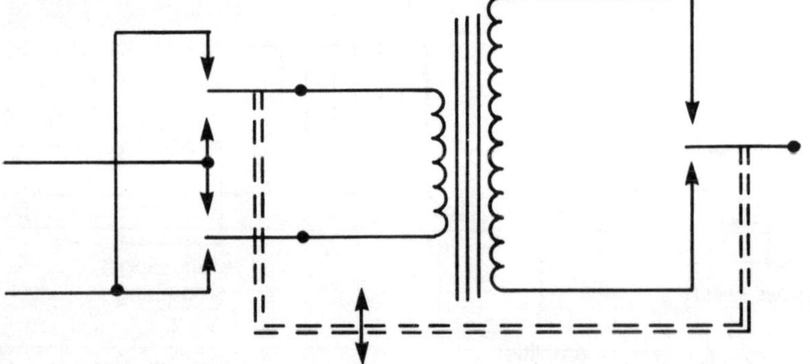

Fig. 3.12 The synchronous switch principle, formerly used for DC-DC converters but now replaced by semiconductor devices.

button used to allow a digital circuit to operate step by step. Another use is to select parameters for the digital circuit, such as the use of DIL switches in devices like computer printers to set up the way that the machine operates. The third method is the switching of digital signals on lines from one unit to another, like the 'T-switches' used to allow one computer to control one of two different printers. This last type of switching can make use of the same types of FET switches as are used for analogue switches, but the way that the switches are used may be different. It is usually undesirable to switch over a digital circuit while the voltages are active, so this type of switch is likely to make use of an *inhibit pin*. While any line is active, a voltage can be generated to keep the inhibit pin at its inhibit level so that the switch will not operate until all the switched lines are at zero level.

Control switches for digital work are generally simple DIL types, usually in banks of eight. Their switching action is on DC voltages only, though the DC voltage does not necessarily control the digital circuit directly. Instead, the circuit interrogates the switch settings at the time when the operating voltage is first applied, and the settings that exist at that instant are used from then on. Changing a switch position in such a digital circuit will not therefore cause any effect until the circuit has been switched off and then on again. This avoids the effects, possibly catastrophic, of unintentional operation of these setting switches while the circuit is active.

The most critical application is for switches that cause a signal input, usually in the form of a push-button switch that delivers a single pulse when pressed. The problem here is that no simple mechanical switch delivers a single voltage change, because the elastic nature of the contacts means that contacts will make, then bounce apart and remake at least once, delivering a multiple pulse. Contact bounce in switches that handle analogue signals will cause noise, mainly impulse noise, but in digital circuits it can cause total failure of the switch action – if the switch is meant to be used to show the digital action step by step, it is hardly very useful if the steps are sometimes single, sometimes double, sometimes triple. Since the bounce is almost unavoidable (though mercury-wetted reed switches are almost bounce-free) the methods used to eliminate the bounce must be electronic.

The main method is the use of an RS-type flip-flop. The characteristics of such a flip-flop are shown in Figure 3.13, and the important point is that the output does not change while the inputs are at the same voltage. In the circuit shown in Figure 3.14, then, the bounce of the

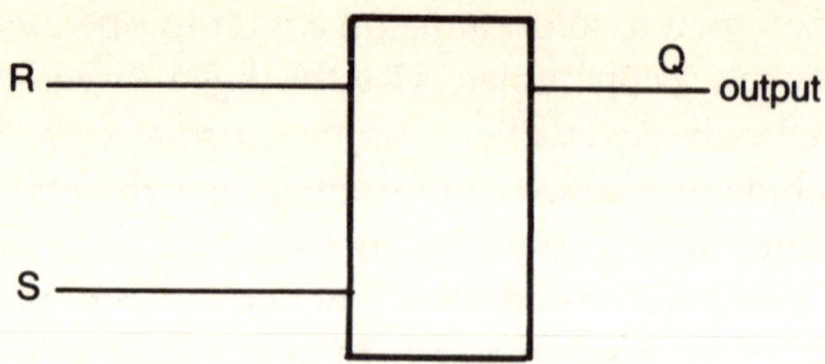

levels: 1=high
0=low

R	S	Q
0	1	1
1	1	1
1	0	0
1	1	0

The input R=1, S=1 is a 'locking' input which preserves the previous condition (0 or 1) at the output

Fig. 3.13 One form of RS flip-flop and its truth table. The important feature for switch debouncing is that the R = 1, S = 1 state maintains the output that was caused by the previous input. Some RS flip-flops use R = 0, S = 0 for this 'locking' state.

switch will only serve to hold the output voltage at its changed level, rather than allowing the voltage to change back. It is unusual to find switches sold along with debouncing circuitry, and the problem of debouncing is usually left to the designer. In some cases (when the switch bounce time is very short) a simple resistor capacitor circuit of the type illustrated in Figure 3.15 can be used. This relies on the bounce time being short compared to the RC time constant, so that the voltage level cannot change by much during the time when the switch contacts bounce open. If a switch that uses such a debounce circuit has to be replaced, then it must be replaced by a switch with identical or very similar bounce characteristics. This is seldom easy, because the bounce characteristics of switches are seldom documented, and specialised equipment (such as a storage or digital oscilloscope) is needed to measure these characteristics.

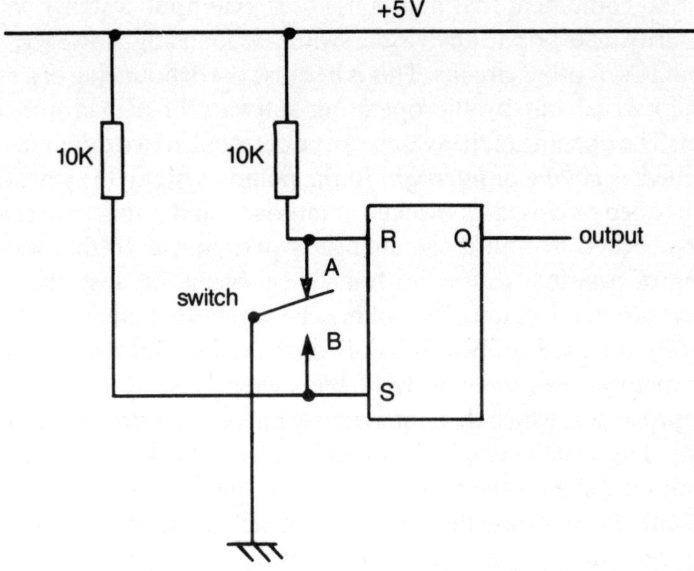

Fig. 3.14 Using an IC form of RS flip-flop to provide a debounced output from a changeover switch.

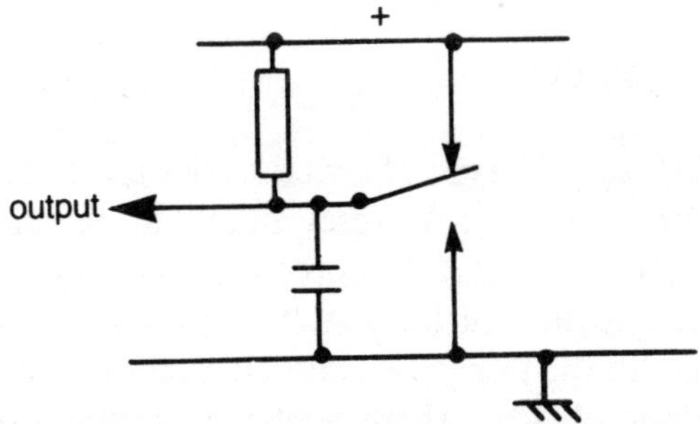

Fig. 3.15 A simple form of debouncing that can sometimes be useful – the capacitor maintains the voltage level with little change during the bounce period. This is useful only when the voltage levels are low, because the presence of the capacitor will cause more contact arcing.

Digital equipment that makes use of signal-input switches will not necessarily incorporate any visible switch debouncing, however, in the form of ICs or other circuits. This is because the debouncing of a switch can be carried out by the operating software of a microprocessor circuit. The opening of the switch can be detected in two different ways, described as *polling* or *interrupt*. In the polling system, the state of the switch (open or closed) is checked at intervals, and if this state changes it is rechecked to find if the change is permanent. In this way, the closing of a switch causes no bounce problems because the switch closure has no effect until the polling check reports that the switch has been found closed on two, possibly more, consecutive readings. The other method uses the closing of the switch to send a signal to the microprocessor, which then interrupts whatever it is processing at that instant. The switch state is then checked, and checked again after an interval longer than the bounce time, and the latter state taken as its final state. In each case the time delay, which is the important feature of any debouncing system, is obtained by a part of the microprocessor software. If problems arise due to insufficient debouncing (such as double letters from a keyboard when a letter key has been struck sharply once), then the remedy is usually a modification to the software, and this may have to be in the form of a *patch*, a piece of added software, rather than a permanent change. The main problems of this type arise with computer keyboards, of which the old TRS-80 Mark 1 was a notorious example, though many modern machines of the PC class are prone to give bounce effects, mainly on the 'n' and 'm' keys.

Dual in line (DIL) switches

The miniature DIL or DIP type of switch is designed specifically for PCB mounting, and is used as a preset mainly in digital equipment, though it is equally well suited for preset applications in analogue circuitry. The main feature of the switch is its very small size. This makes operation with the finger very difficult, so the usual method is to use a screwdriver blade. These switches should preferably be mounted where a slip with the screwdriver blade will cause the least possible damage, electrical or mechanical, to the rest of the circuit. The usual formats are four-pole or eight-pole on/off, and the action is a slide or a lever toggle. A changeover action can be obtained on suitable types (in which one lever operates two switches) by connecting a pair of switch pins together. DIL switches are not intended for continual use, and have a comparatively low life expectancy of around 20,000 operations. To put this in perspective, you might expect to alter the settings once or twice a year, at the most once a week.

One of the main problems attending DIL switches is mechanical damage. Designers have an unfortunate knack of placing these switches where they are difficult to see and even more difficult to operate. One printer mechanism, for example, requires the casing of the printer to be removed simply in order to set the line-feed switch, and many users of this printer cut a hole in the casing above the switch so it can be operated without dismantling the machine. This, however, means that the switch is being operated by a screwdriver stuck through a hole in the casing, which is the main source of damage to the switch and to surrounding circuitry. It is not usually possible to make amends for inconsiderate design, but in some instances the switch can be replaced by a miniature socket and the switch relocated at the end of a cable.

Keypads

The keypad type of switch, used either as the small numerical keypad or as the full-blown alphanumeric computer keyboard, is usually mechanically simple and requires debouncing circuitry or software. The switch is usually arranged so that contact is made about halfway along the travel of the key, and a spring is used to impart some 'feel', to the keyboard and to prevent the contacts from being pushed together with excessive force by the sort of typing that is more thump than touch in nature. Just as the mechanical quality of the switches can vary from the simple phosphor-bronze leaves to the Hall-effect magnetic, so the electronics of the keyboard can range from simple matrix to the completely debounced and decoded. The basic keyboard, however, is always connected using on/off switches in a matrix pattern.

The basis of a matrix keyboard is illustrated in Figure 3.16. The keys are arranged in lines and columns, and each switch will have one connection made to a line and the other to a column. This greatly reduces the cabling to the keyboard, because an array of 64 switches in an 8 × 8 matrix needs only 16 connections. The matrix arrangement,

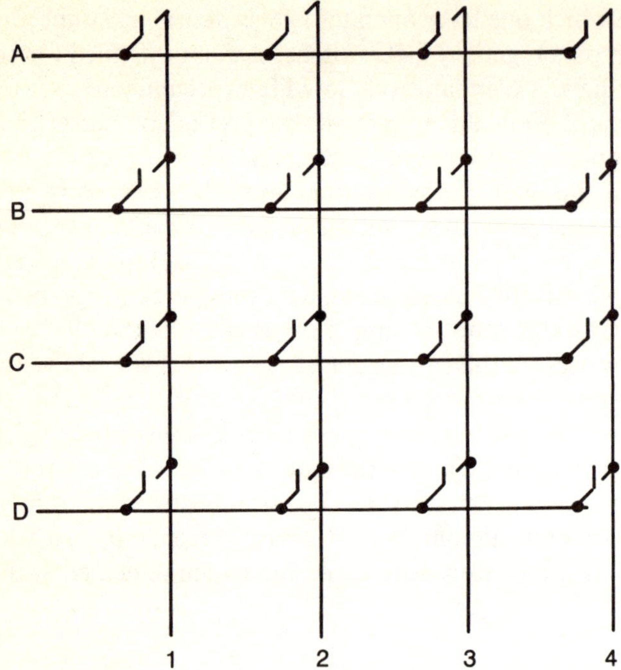

Fig. 3.16 The matrix arrangement of switches on a typical keypad or keyboard.

incidentally, applies only to the electrical connections – the physical arrangement of the keys on the keyboard can be anything that is required by the designer, though very few keyboards depart from the calculator or telephone pad style for numeric, or the typewriter arrangement for alphanumeric use. Because of the matrix connection, the closure of a key does not have the effect of a conventional switch in causing connection between two points that are unique to that switch. In a switch matrix, each row or column wire will be used by more than one switch, and the closure of one switch will cause a connection between one row and one column, (see Figure 3.17). This particular combination will often be unique to that switch, though on some arrangements the simultaneous pressing of two keys can produce a false output – on my PC keyboard, careless pressing of OM can produce OM7, for example. A matrix keyboard or keypad is therefore of little use in its natural state and needs to be decoded, in the sense that combinations of connections between rows and columns have to be translated into signals that are related to the keys.

The decoding can be carried out in the keyboard itself, using a

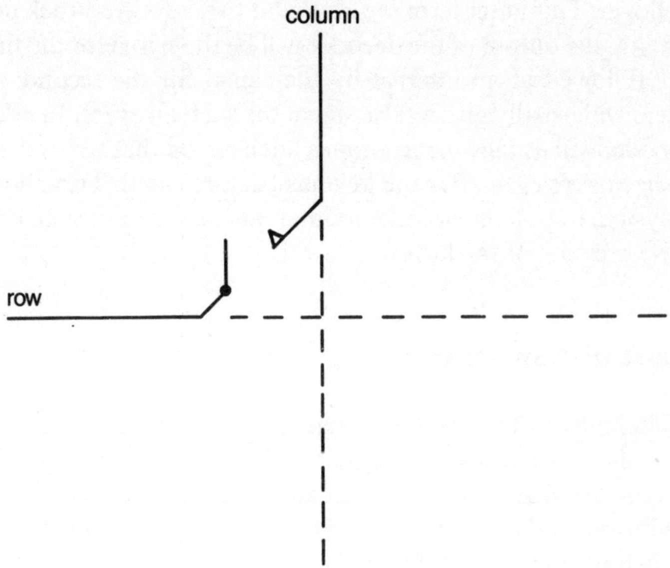

Fig. 3.17 Each switch makes a contact between one column and one row. The matrix of rows and columns must be connected to a decoder to convert each key-press into a signal on a separate line.

matrix IC, and when this is done the keyboard is sometimes described as *decoded*, *intelligent* or *smart*. However, the use of such keyboards is declining because the connections to a keyboard that has been decoded must consist of at least one lead per key. Now that keyboards use at least 80 keys, this makes for rather bulky cables between the keyboard and the rest of the equipment. The situation is not so serious for keypads, most of which use only 16 keys, but it is very unusual to find a decoded keypad. The normal solution is to connect the keyboard or keypad to the main unit using only the row and column connections, and to decode in the main unit. When the keyboard or keypad is part of a computer system, the decoding will invariably be carried out by software, and this software will include the key debouncing circuits and provision for making the key action repeat if the key is held down. Where a keypad is used in equipment that does not include a microprocessor (now rapidly becoming an endangered species) a decoder IC can be used. The use of such ICs provides much more than simple key detection, and it would be difficult to justify the use of any other method of decoding. Just to give one example, even a low-cost decoder could be expected to provide effective debouncing and also key rollover. This latter term means that if two keys are struck in rapid succession, the output of the decoder will be the signal for the first key struck, followed at an interval by the signal for the second; simple decoders will usually give a false signal for such an event. In addition, the decoder will usually incorporate a latch circuit that stores the result of a key closure even after the key has been released. This allows the main system to poll the decoder at intervals that are not synchronized with the pressing of the keys.

VHF and UHF switching

The switching of VHF and UHF signals presents problems that do not arise in the consideration of other switches. The frequency of the signals implies that even very small amounts of stray capacitance can be significant, and the inductance of the switch components can lead to the switch having a resonant frequency which may be in the VHF or UHF range. In addition, the switch will normally be located in a line whose impedance will be of the order of 50 Ω, and the switch should also have this impedance value in order to avoid mismatching on the line. Very simple designs exist and are used, for example, for switching the input to a TV receiver between an aerial source and a TV game source. These switches are not intended for anything other than casual domestic use, however, and have large losses. The better quality switches for this frequency range are of coaxial construction so that matching can be maintained, and the operation is more likely to be by solenoid. This permits remote control, so avoiding hand capacitance effects.

For switching signals at low level that are not on a line, mechanical switching is seldom useful, and has been replaced by the use of low-capacitance diodes, usually of the Schottky type. The bias on these diodes is taken through a well decoupled connection to a remote switch, so that the main switching action is simply a low-voltage DC one. The design of switching for UHF/VHF signals at high power levels is a very specialised matter and usually calls for a one-off solution rather than any form of off-the-shelf switchgear.

Chapter Four
Other Mechanisms

The switch mechanisms we have examined so far are the conventional types that account for more than 90% of all switch sales. There are, however, many important types of switch that are not sold in large numbers. These include switches with conventional contacts which are operated by mechanisms other than the conventional types discussed above, and also switches in which the connection is made by other than conventional contacts. The main operating method considered so far has been the human hand, but the use of switches in automation implies operation by other mechanical methods which can be of considerable importance. Nor have we mentioned switches operated by light, temperature change, acceleration and other physical factors. Some switches operated by physical factors are better classed as sensors rather than switches, but if their action is to make or break an electrical circuit they will be dealt with in this chapter.

Operating mechanisms

An operating mechanism frequently used in machinery is the *cam*. For the type of switching that we are considering, cam operation will be applied mainly to microswitches, and the small motion needed to operate a microswitch makes the cams easier to cut. In addition, the use of a microswitch makes it easier to use cams with many lobes, and of light and simple construction. The main point to consider is that the operating lever of the microswitch will absorb any excess movement so that the microswitch itself is not damaged. The provision of cams and their bearings, together with mountings for the microswitch, is a matter for the designer of the equipment, and a microswitch fitted with a roller leaf actuator should always be specified.

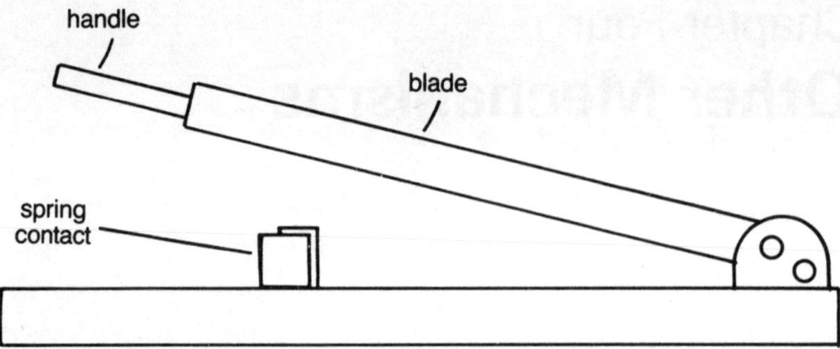

Fig. 4.1 The form of the old-style knife switch; this is still found in some applications for low-voltage high-current work, such as electroplating.

A switch mechanism that received no mention earlier is the *lever* type, illustrated in Figure 4.1. This type of switch was originally a Post Office design (in the days when the telephone service was part of the Post Office) and is rather a rare item nowadays. The centre position is the neutral position, and movement to either side operates changeover contacts, usually four on each side. The contact specification calls for the use of silver, rated at 100 V 300 mA DC, or up to 250 V 300 mA AC. This type of switch is useful for (mainly) DC and low-frequency signal circuits in which a large number of changeover actions are needed.

Joystick action is a comparatively recent development, and one which is being used increasingly in control applications. Joysticks are also used in combination with potentiometers, but the *switching* type will generally be used, and these recognise the direction in which the joystick is pushed by employing four switches. In some examples these are four standard-pattern changeover microswitches, and the principle used is that the four main directions (labelled as north, south, east and west) will each cause a single microswitch to change over, while the mid-directions of north-east, south-east, south-west and north-west will each cause two switches to change over. *Wafer-switch* joysticks are also available: typically these employ two four-pole three-way switches. Whatever type is used, the centre position is the 'off' position for all switches. Joystick control is used along with direction-specifying equipment in robotics and machine control, and although the analogue type of joystick (using potentiometers) allows closer control, the switch type has the merits of simplicity and long life.

Foot operation

The operation of switches by the foot has a large number of applications in control circuitry, so that foot-operated switches are a feature of most components catalogues. A foot-operated switch is likely to come in for harder and less sensitive use than a hand-operated switch, and the mounting and operating action of the switch are important features of the design. The enclosure must be robust and also reasonably heavy, so that it cannot easily be kicked aside. It must also grip the floor surface well so that an attempt to operate the switch does not simply result in the unit sliding along the floor. The pad which is in contact with the foot also has to be constructed from non-slip material, and the assembly should be waterproof. The switches used generally employ SPDT action, since foot operation seldom calls for much more than on/off or changeover action.

Air actuators

Pneumatic switch actuators are an old-established method of operation that still have a large number of uses. Air actuation, for example, can be a useful option in flammable atmospheres, with the switch situated remotely in a safer environment. Another attraction is that the air-pressure switch can be of a standard design, with only the pneumatic transmitting elements differing according to whether the switch is to be operated by hand, foot, or some other limb movement, or by force on a plate, depth of liquid or any other source of pressure. The snag (as compared with remote electrical actions) is that the air in the tubing of a pneumatic actuator will expand with increasing temperature, so that the switch can be operated by temperature alone. This means a switch could be turned on by high temperature, or become impossible to turn on because of low temperature. Since the change of pressure caused by change of temperature is also affected by the amount (mass) of air in the system, false operation can be avoided by keeping the length of connecting tubing below a stated maximum, depending on the type of actuator and the temperature range required. These critical lengths range from just over a metre to some 30 metres, and would be important if pneumatic operation were being used in order to avoid having current make and break contacts in a flammable atmosphere.

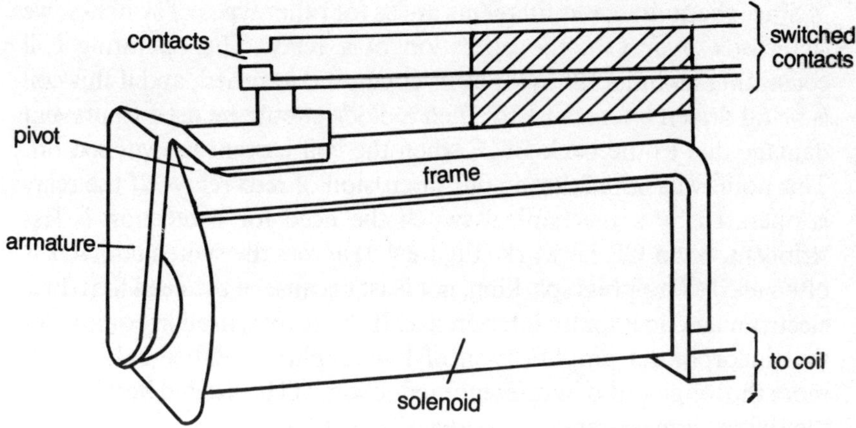

Fig. 4.2 The classic type 8000 relay construction.

Magnetic operation

Magnetic operation of switches has been dealt with to some extent in the previous chapter, in connection with microswitches. In general a magnetically operated switch is classed as a relay or contactor, and detailed treatment is beyond the scope of this book, but since the contact principles for small relays are identical to those of mechanically operated switches it would be illogical to exclude relays altogether. In addition, there are other magnetically operated devices that need to be dealt with under this heading.

The classical form of relay is illustrated in Figure 4.2. The coil, usually a solenoid, is wound on a former around a core, and a moving armature forms part of the magnetic circuit. When the coil is energized, the armature moves against the opposition of a spring (or the elasticity of a set of leaves carrying the electrical contacts) so as to complete the magnetic circuit. This movement is transmitted to the switch contacts through a non-conducting bar or rod, so that the contacts close, open or change over depending on the design of the relay. The contacts are subject to the same limits as those of mechanical switches, but the magnetic operation imposes its own problems of make and break time, contact force and power dissipation. Like so many other electronics components, relays for electronics use have been manufactured in decreasing sizes, and it is even possible to buy relays that are packaged inside a standard TO-5 transistor can.

Since the contact requirements are as for other types of switches, we shall look mainly at the operation of a relay. The operating coil constitutes an inductor as far as its circuit is concerned, and if this coil is being driven by a transistor then a diode should be used to prevent damage due to the back-EMF when the coil current is switched off. This point was noted during our discussion of reed relays. If the relay is operated by a mechanical switch the need for protection is less stringent, but a CR network (Figure 4.3) across the switch contacts is often used to suppress sparking, not least because of the need to reduce electromagnetic impulse interference. If the relay is used in equipment that incorporates amplification of low-amplitude RF signals, much more thorough radio interference reduction will be needed both across the switch contacts and across the relay contacts.

The specification for a relay will include the permitted range of current or, more commonly, voltage that can be applied to the coil. At the minimum level and below, correct operation cannot be ensured. The relay may hold in at a voltage or current considerably lower than the specified minimum, but cannot pull in its armature if the relay has been off. The maximum rating is due to dissipation, so that exceeding the maximum voltage or current rating will cause overheating, and will also cause more severe contact bounce. The normal rated values that are quoted are usually close to the lower limits, so that a typical percentage tolerance is –8% +60%. Failure to maintain correct voltage on a line used to energise a relay can therefore lead to incorrect operation if the voltage falls by a comparatively small amount. For

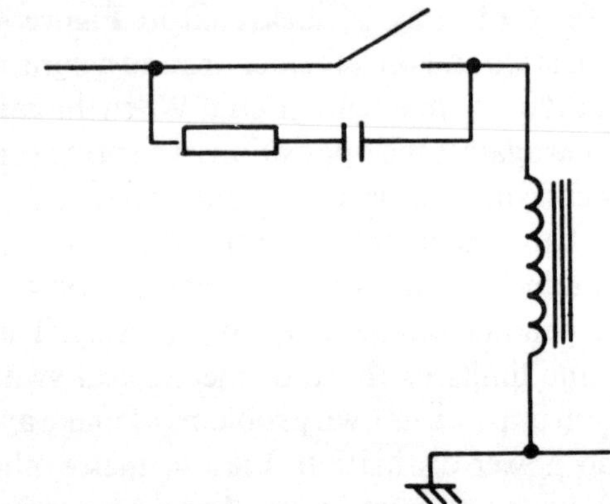

Fig. 4.3 Using CR arcing suppression on switch contacts that operate a relay coil.

many purposes, using a large reservoir capacitor on such a supply eases problems of maintaining voltage, because if the capacitor can maintain the voltage level for long enough (typically 0.5–10 ms for small relays) then a drop in voltage of more than 8% will be harmless, since this will still be enough to hold the relay.

The use of a relay as a remote electrically controlled switch is just one aspect of relay use. Relays are used where the use of switches alone would require impossible mechanical interconnections or awkward electrical cabling. They can be placed where a mechanical type of switch could not be reached, in environments where sealed contacts must be used, or where the contacts have to handle voltages or currents that are beyond the range of normal switches. An important feature of almost any type of relay is its electrical isolation between the operating coil and the contacts. For some uses, safety considerations demand the use of relays when a mains (or higher) voltage supply has to be switched by low-voltage equipment with no earth connection. Relays are also used when a sequence of switching has to be carried out, initiated by making contacts on one single switch, or where a safety

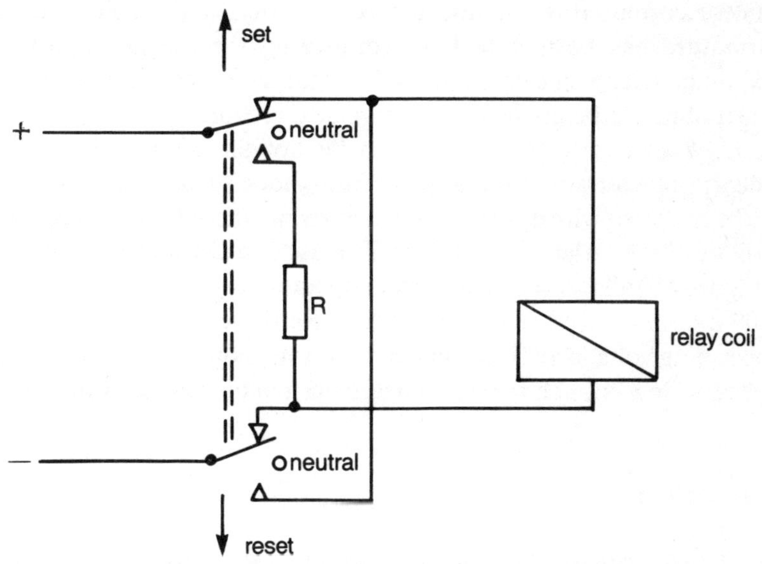

Fig. 4.4 Switching a remanence relay. The relay will remain in its on or off state with no power applied.

cutout is required that will provide isolation from high voltages until reset.

Like switches, relays also exist in special forms. In the *latching relay*, a mechanical tooth and ratchet will maintain the contacts in a set position. The normal form of relay operates its contacts for as long as current flows in the coil, but a latching relay can be used to hold contacts either made or broken, reversing the action each time current is applied to the coil. A similar action can be obtained without any mechanical complications using a *remanence relay*, in which the coil current is applied for a controlled time (typically more than 10 ms and less than one minute) This magnetises a core made from an alloy that retains its magnetism after the coil current is switched off. The retained magnetism then holds the armature in place until a reverse current is passed through the coil. The reverse current is usually smaller than the forward current, and a circuit arrangement such as that in Figure 4.4 is used with a resistor to control the reverse current. A third method of achieving a latching action is to incorporate a permanent magnet into the core path. If the strength of this magnet is midway between the holding and pull-in levels, then such a magnet cannot operate the relay unless current flows in the correct polarity in the coil. Once the armature has been pulled in, completing the magnetic path, the permanent magnet can maintain this state. To release the relay a lower current has to be applied in the reverse direction.

High sensitivity relays can also be bought off the shelf. These incorporate a transistor and protection diode so as to allow the relay to be operated with a very low current swing (though the voltage swing will be of the order of 0.6–1.2 V). The use of a Darlington pair circuit (Figure 4.5) allows very low current operation, typically of the order of 100 μA, to be used. The use of transistor drivers can ensure that the operating voltage and current can be at safe, non-sparking levels, with currents low enough to be no hazard even in biomedical applications.

Hall effect

Hall-effect devices form an entirely separate class of magnetically actuated switches that have important uses of a very different nature. The basis of the Hall effect is the force that acts on any conductor which carries current while situated in a magnetic field (see Figure 4.6). This force is exerted on the particles that carry the current rather than

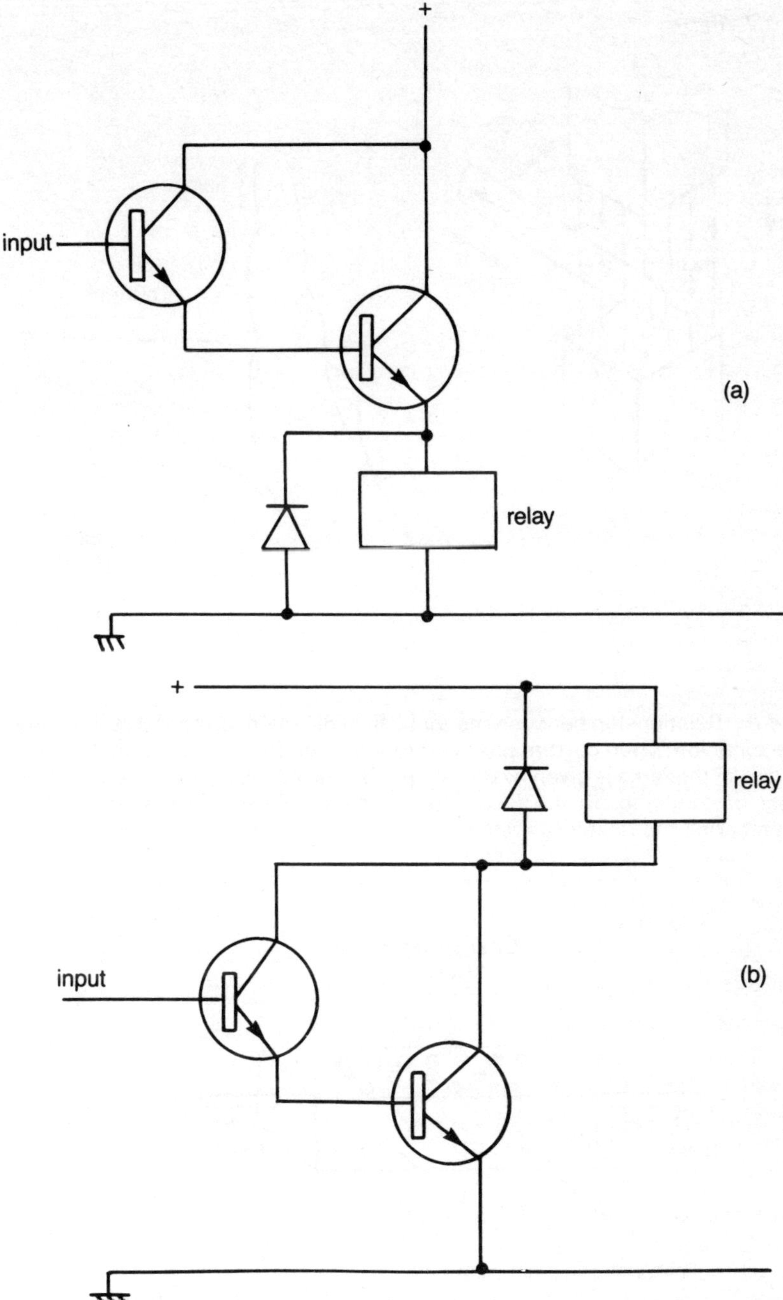

Fig. 4.5 Driving a relay with a Darlington circuit (a) Emitter loading, which provides large current sensitivity but requires the input signal to change voltage by rather more than would normally be needed to operate the relay. (b) Collector loading, providing both current and voltage sensitivity. The change of voltage at the input need only be enough to switch the transistors on.

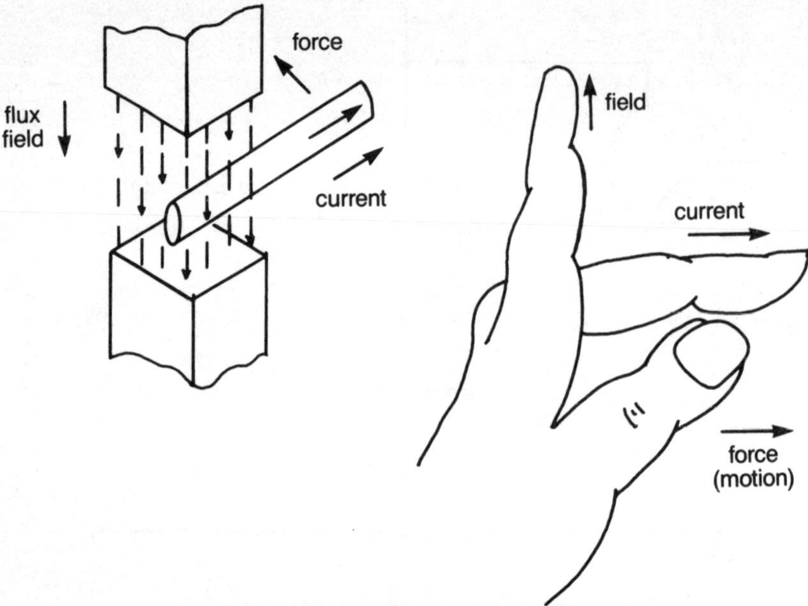

Fig. 4.6 Relationship between magnetic field direction, current direction and force direction when current passes through a conductor in a magnetic field. The size of the force is given by B.I.l, where B is flux density, I is current and l is length of conductor in the field. The left-hand rule is a useful method of remembering the relative directions.

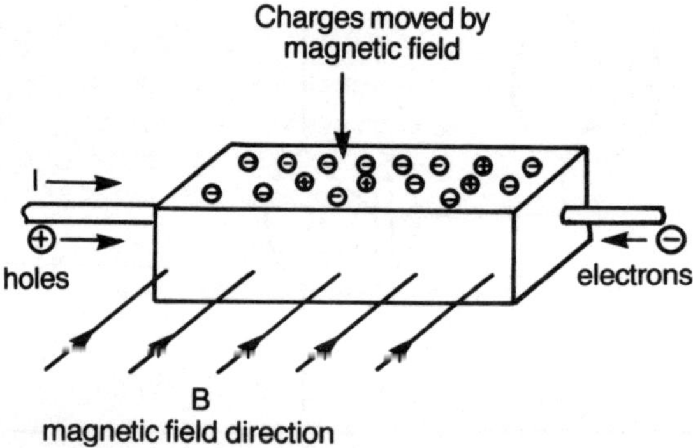

Fig. 4.7 The Hall effect. The effect of the magnetic field on a current carrier is to deflect the carrier in the direction shown. The polarity of the voltage across the faces depends on whether the majority carriers are electrons or holes, but the size of the voltage is very closely proportional to magnetic field strength if the current is constant.

on the fixed atoms of the conductor, so that each moving particle in the conductor has a force acting on it that will deflect its direction of movement, as shown in Figure 4.7. In good conductors, such as metals, the predominant charged moving particles are electrons, and because the number of electrons that are free to move is very large, the actual average speed is very low – a few centimetres per hour. This very low speed makes the force exerted by a magnetic field very small, so that the effect is extremely difficult to measure in such conductors. It was, however, discovered and reported by Hall at the end of the nineteenth century, a long time before semiconductor materials were available.

A semiconductor has a much lower concentration of charge carriers, and to carry a given current these must move much faster than their counterparts in a metallic conductor. This makes the force exerted on each particle in a magnetic field correspondingly greater, and the result is that particles are detectably deflected. As Figure 4.7 shows, if we imagine current carriers moving in a rectangular slab, the effect of a magnetic field acting in a direction at right angles to the direction of the current will be to make one face of the slab more positive than the opposite face. This constitutes a voltage, called the *Hall voltage*, and for a given sample of semiconductor this Hall voltage is proportional to the strength of the magnetic field. The voltage is low, of the order of millivolts in a field produced by a permanent magnet, but since the Hall effect can be produced in a semiconductor strip that is part of an IC amplifier there is no problem in obtaining outputs that are adequate to operate other devices. In particular, the IC associated with the Hall-effect detector can incorporate a switchover circuit (a Schmitt trigger) that ensures a very rapid change between states at a critical level of magnetic field, so that the output is a switching output rather than an analogue one.

The Hall effect therefore provides a method of switching that is totally free of switch bounce, and as such is used in high-grade keyboards. The operation of any Hall-effect device depends on the use of a magnet, and in a Hall-effect key switch the semiconductor slab will be fixed in the key frame and a permanent magnet will be secured to the key. Apart from keyboard use, the Hall-effect switching device is used extensively as a proximity switch along with a permanent magnet. The switchover distance is typically a few millimetres in such an application, though it can be increased by mounting the detector part on ferromagnetic material so as to form a more efficient magnetic circuit. Operation of a Hall-effect device by an electromagnet is less usual, but

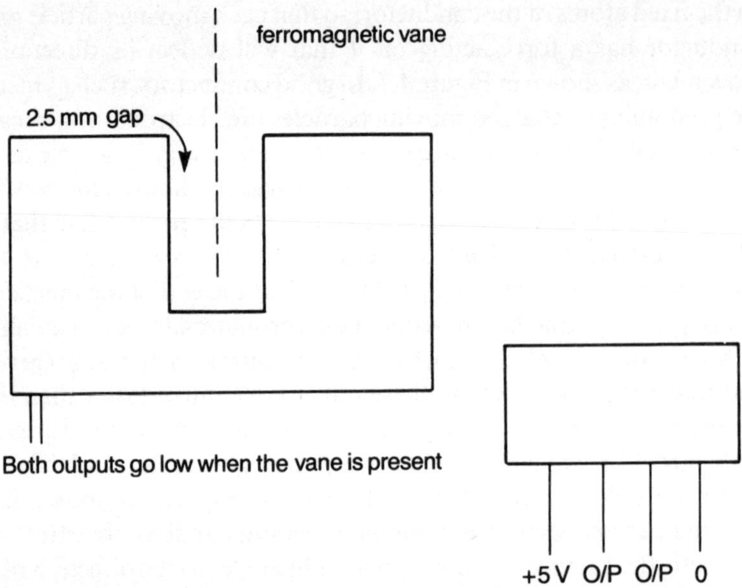

Fig. 4.8 The form of a Hall-effect proximity detector. This gives an output signal when a ferromagnetic material passes across the slot.

this can form the basis of an excellent bounce-free, contactless relay ideally suited for use in environments that are impossible for conventional relays. A suitable combination is the Hall-effect switch chip with the operating coil for a reed relay, and these two can be encapsulated in resin to form a unit if needed.

The Hall effect is also employed in switch units that will sense the presence of a ferromagnetic vane between two surfaces (Figure 4.8). A unit of this type can be used in situations where the conventional method of counting objects by interrupting a light beam is unsuitable. In particular, the Hall-effect device is unaffected by bright surrounding light or by the presence of dust or smoke.

Semiconductor switching

The Hall effect has reintroduced the topic of switching by means of semiconductors which was briefly dealt with in the previous chapter. In this section we shall not be concerned with the switching effect of a single bipolar or field-effect transistor but with the Schmitt trigger and

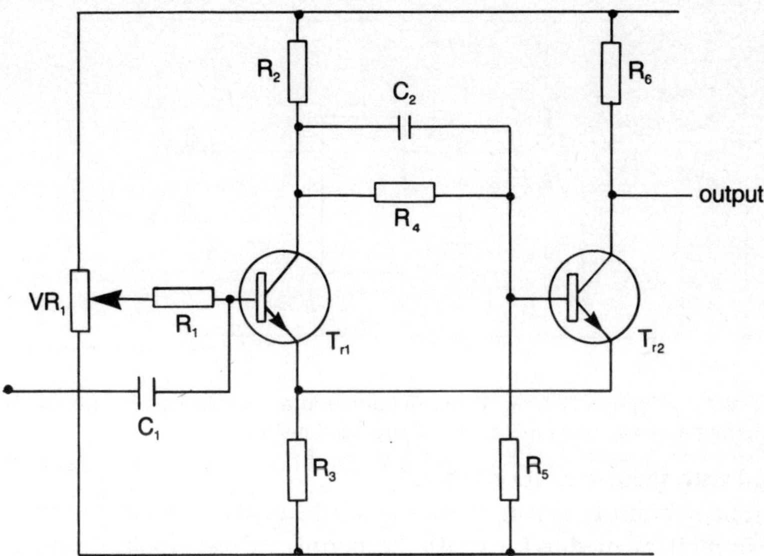

Fig. 4.9 The discrete-transistor form of a Schmitt trigger. The circuit is nowadays more often used in its IC form.

the four-layer semiconductor switching devices such as the thyristor. All of these are electrically operated switches, and so are not the primary concern of this book, but since their use is so widespread, and so often occurs in conjunction with conventional switches, we must

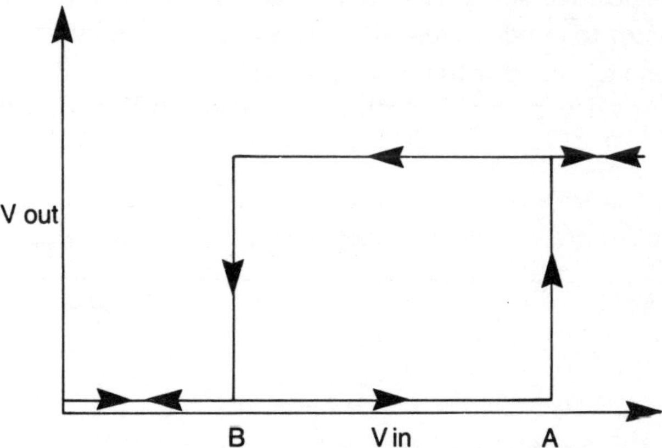

Fig. 4.10 The input-output characteristics of a Schmitt trigger. A rising input voltage has no effect until its level reaches point A, when the output switches to a high level. When the input is lowered again, there is no effect until the voltage level reaches B. The switch-over can be very rapid, typically 20 ns for IC versions.

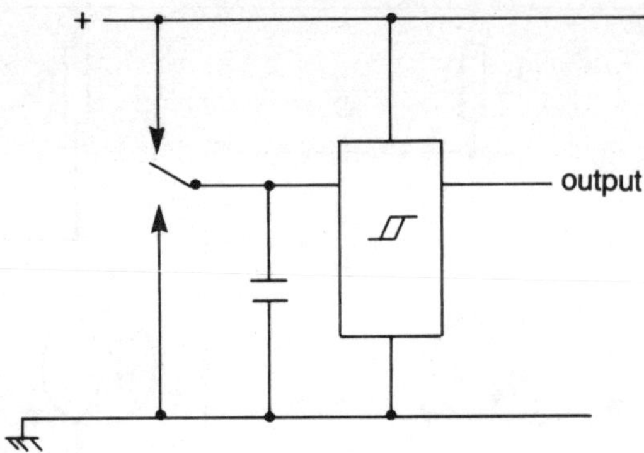

Fig. 4.11 Using a Schmitt IC for debouncing a switch. This is often a more satisfactory method than the use of the RS flip-flop.

deal with them even if only briefly.

The Schmitt trigger is an old-established switching circuit that has been implemented in turn with thermionic valves, bipolar transistors and field effect transistors. It is still the basis of many switching devices in IC form. A simple version, using bipolar transistors, is illustrated in Figure 4.9. The normal circuit bias is arranged so that Tr2 is normally conducting, and the current that is passed through the common emitter resistor R3 is sufficient to keep Tr1 biased off. In many applications the DC voltage of the base of Tr1 will be earth, but the sensitivity of the circuit can be increased by biasing this base slightly positive. A positive-going signal at the base of Tr1 will eventually cause this transistor to conduct, and when this happens the fall in voltage at the base of Tr2, along with the drop in current through R3, will make the switch-over very rapid. Within a microsecond or so, Tr1 will be

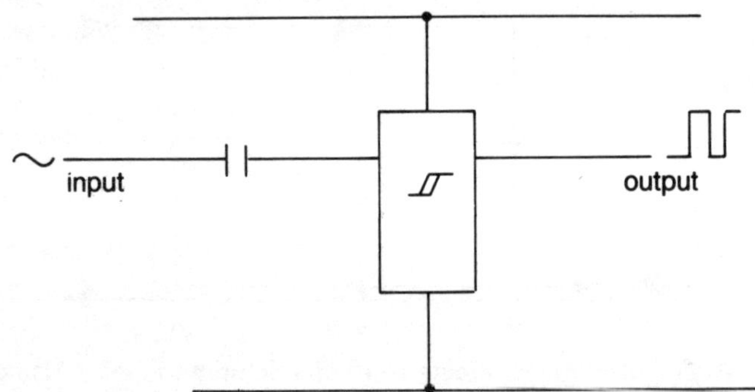

Fig. 4.12 Using a Schmitt IC to convert from a sine wave to a square wave with very sharp rise and fall.

fully conducting and Tr2 shut off. This persists only for as long as the base voltage of Tr1 causes conduction, however, and when this base voltage reaches a lower level, the rise in collector voltage of Tr1 will allow Tr2 to conduct, causing the circuit to snap back to its original state. Once again the change is very rapid, of the order of a microsecond in the version illustrated and very much faster in TTL IC circuits.

The important feature of the Schmitt trigger, apart from its very fast contactless switching, is the difference in voltage levels between switch-on and switch-off. This difference is called *voltage hysteresis*, and it can be illustrated in a diagram such as that in Figure 4.10, which shows the higher voltage at which the device switches over when an input voltage is rising, and the lower level at which it switches back. The difference between these levels serves to make the device bounceless, and one common use of the IC form of the Schmitt trigger is to provide debouncing for a mechanical switch, as illustrated in Figure 4.11. The Schmitt trigger can be used to advantage wherever a slowly changing or analogue voltage has to be used to generate a switch action, usually as part of a digital device. One example is in deriving trigger pulses from sine waves, using a circuit of the type outlined in Figure 4.12, but a more common application in a true switching application is to generate a switching output from the slowly changing inputs of sensors such as thermistors, Hall-effect devices, liquid level detectors, photo-cells and similar analogue devices.

An entirely different class of semiconductor device makes use of a single semiconductor crystal with four separate layers, as shown in Figure 4.13. The best known device of this class is the *thyristor*, whose action is very similar to that of the gas-filled thyratron device of earlier years. The thyristor uses connections to only three of the four semiconductor layers, referred to as *anode*, *cathode* and *gate* respectively (Figure 4.14). The voltage to be switched is applied between the anode and cathode, with the anode positive with respect to the cathode. With the gate at cathode potential, no current will flow between anode and cathode, but when the gate voltage rises to a given level (usually above 0.6 V) then conduction between anode and cathode is triggered. From then on it will continue until the voltage between anode and cathode falls to a very low value, around 0.2 V. The gate can be triggered by a pulse provided that the gate voltage and current are held at triggering level for a sufficient time, around a microsecond. The switch-on action is similar to that of a latching type

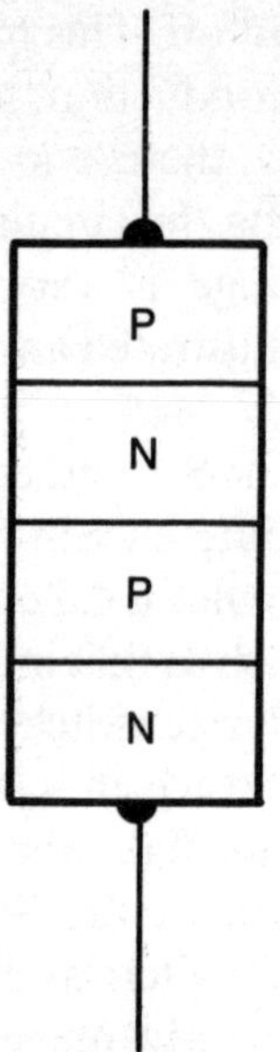

Fig. 4.13 A four-layer semiconductor. The characteristics of a device formed in this way depend on what connections are made to the intermediate layers.

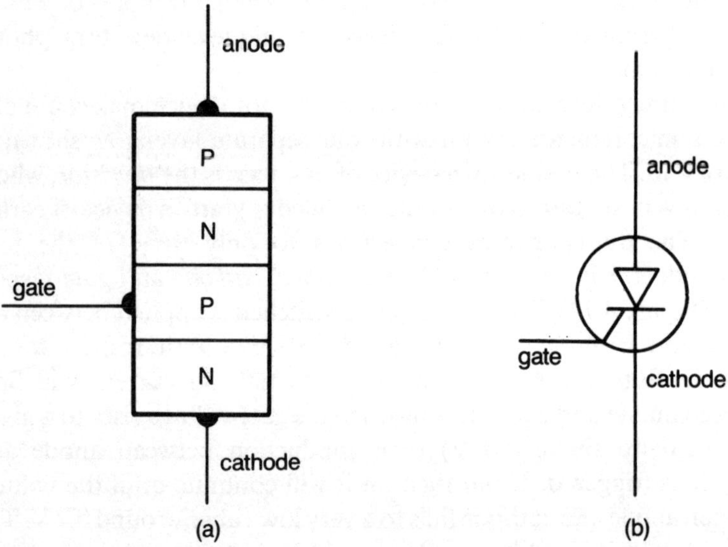

Fig. 4.14 The thyristor, showing layer connections (a) and conventional symbol (b). Some old texts use the name 'silicon controlled rectifier'.

of relay, but there are important differences. The power required for switching at the gate is very low. The time is very short compared to that required for a mechanical device (though long compared to other semiconductor switches). The device is unidirectional (current will not flow unless the anode is positive relative to the cathode). Finally, there is no method of turning off conduction other than reducing the voltage and current between anode and cathode to less than the critical holding level (but see gate turn-off devices, below). Note that the current requirements at the gate call for low-impedance circuits, since gate currents of 100 mA or so may be needed for a brief period.

The thyristor is used in relay-type circuits, particularly where the voltage to be switched is AC. This automatically provides for turn-off when the voltage falls to zero in the middle of each cycle, but in DC circuits turn-off can be arranged by shorting the supply briefly with a second thyristor. The thyristor can pass large currents and has the advantage of being contactless, unlike electromagnetic relays, but it has the disadvantage of having no isolation between the gate circuit and the circuit that is being switched. This arises because the gate voltage must be applied between gate and cathode. In many circuits, then, a pulse is fed to the gate-cathode circuit by means of a

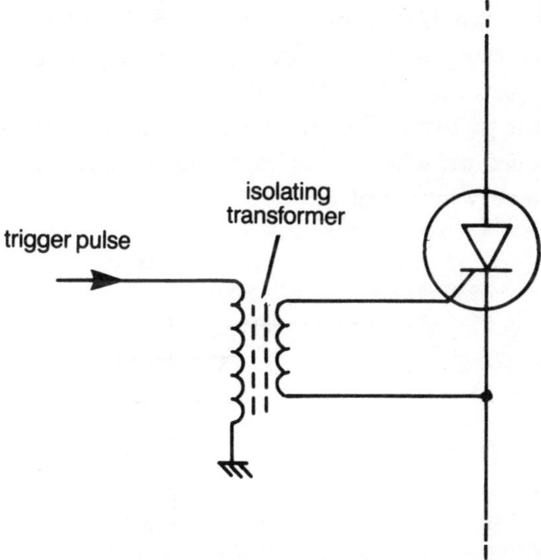

Fig. 4.15 Operating a thyristor through a pulse transformer to isolate the triggering circuit from the switched circuit. Since only a brief pulse is required, the transformer can be a miniature type, often wound on a ferrite ring.

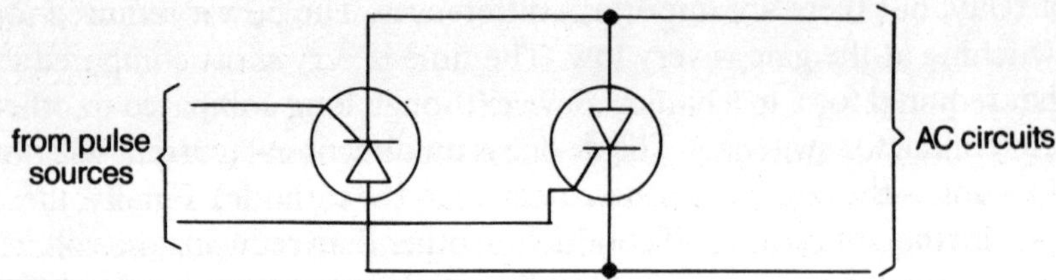

Fig. 4.16 Full-wave AC switching using two thyristors back-to-back.

transformer, which can have very high resistance and breakdown voltage ratings between primary and secondary (Figure 4.15). Another triggering option is the *opto-isolator*, dealt with later in this section. Some equipment specifications will not, however, accept the use of thyristors in place of electromagnetic relays, particularly in domestic appliances that are subject to high temperatures (such as central heating controllers).

For controlling the full wave of AC, circuits that use two thyristors connected anode to cathode (Figure 4.16) can be used, but a more common solution involves the use of a bidirectional device known as a *Triac* (the name is a trade mark), as shown in Figure 4.17. The electrodes of the Triac are labelled as MT1, MT2 and gate, with the voltage to be switched applied between MT1 and MT2 in either polarity. Triggering is also bidirectional for a swing of about 2 V between the gate and MT1, though the triggering current required depends on the polarity of the triggering signal. Like the thyristor, the Triac is switched off when the voltage and current between MT1 and MT2 are at a very low level.

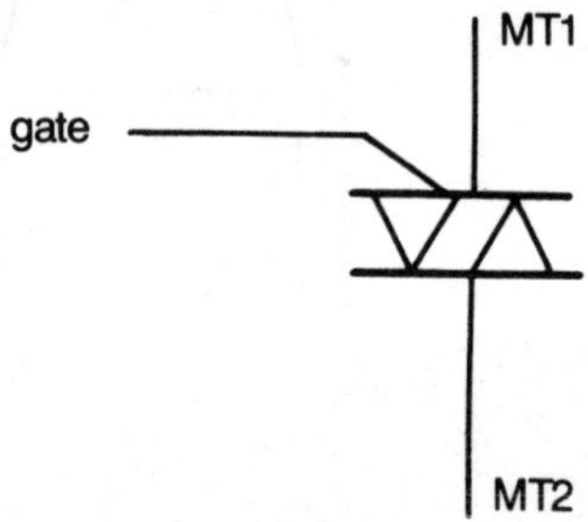

4.17 The Triac symbol. This device conducts between the MT1 and MT2 terminals when triggered by a pulse at the gate. The pulse can be of either polarity, but the triggering amplitude for a negative pulse is not the same as that for a positive pulse.

There is one thyristor-like device that can be turned off electrically, known as the *Gate Turn-Off* (GTO) thyristor or switch. Typically, a device of this type will permit conduction between anode and cathode for a positive pulse of specified voltage, current and duration, and will turn off with a negative pulse of larger voltage but shorter duration. Other devices of the four-layer type include the *Silicon-Controlled Switch* (SCS), a device in which a lead is taken to each of the four semiconductor layers. This allows the SCS to be used in two modes, either as a form of unijunction transistor or as a thyristor.

Other mechanical switches

For specialised purposes, a variety of mechanical or electromechanical switch devices exist that switch over because of some effect other than manual action or the presence of a magnetic field. Some are purely mechanical, but some depend on actions such as that of the Schmitt trigger (described above) to achieve a switching type of output. Some of these devices can be obtained in off-the-shelf form, others have to be made up as required from standard components. Most are subject to some form of adjustment when installed to make sure that their action is suitable.

The *mercury tilt switch* is an example of this type of device. The principle (Figure 4.18) is that two metal contacts are embedded in a curved glass tube that also contains the liquid metal mercury. As the tube is tilted, the mercury moves until it bridges the electrodes and makes electrical contact. The switching is bounce-free and clean even if the tilt speed is slow, so that no form of Schmitt trigger is needed.

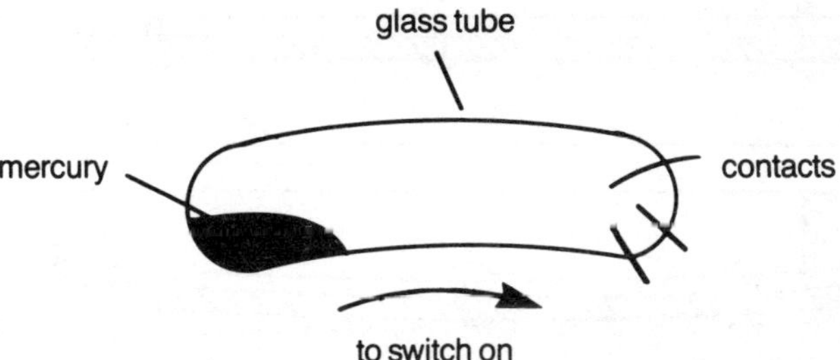

Fig. 4.18 Mercury switch principle. The tube is banana-shaped, and the switch is operated by tilting the tube so that the mercury flows towards or away from the contacts.

Since the switch is a sealed assembly that can be obtained completely encapsulated it can be used in hazardous atmospheres, and since no conventional mechanical contacts are used, the life of the switch is very long. The current rating depends on the size of the internal electrodes, and is generally a fraction of an amp for miniature devices, several amps for the larger types. If very high current levels are to be switched, the tilt switch should be used to drive a large contactor, because there is a danger of very large currents vaporising the mercury. If this led to a fracture of the container, the spread of toxic mercury vapour would constitute a considerable hazard to anyone nearby. For the same reason, mercury tilt switches, mercury-wetted reed relays and ordinary mercury thermometers should never be cut open – a precaution that brings smiles to generations of science teachers who used to pour mercury about with considerable abandon. Note, incidentally, that mercury freezes at that very memorable temperature which is the same on both Fahrenheit and Celsius scales, –40°.

Thermal operation of switches is important for detection of effects as diverse as fire, overheating and the failure of a freezer. The simplest type of thermal switch is the *bimetallic* type, whose principle is illustrated in Figure 4.19. A strip is formed with two layers of different metals, chosen so as to have very different expansivities. The *expansivity* (the current term for what used to be called the *coefficient of expansion*) is the fractional change of length per degree change of temperature. Since this is positive for all metals, the strip expands as the temperature increases. Because the metals do not expand by the

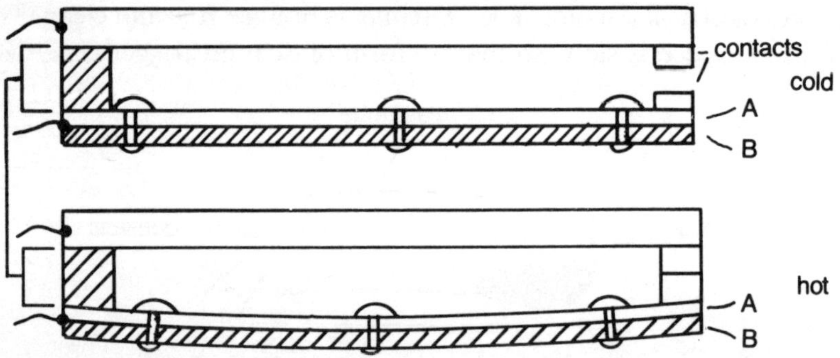

Fig. 4.19 Bimetallic strip principle. The metal B has a higher expansivity than metal A, so that a rise of temperature causes the strip to bend, making contact. The contacts may be arranged so that connection is broken as temperature rises (as for a domestic thermostat).

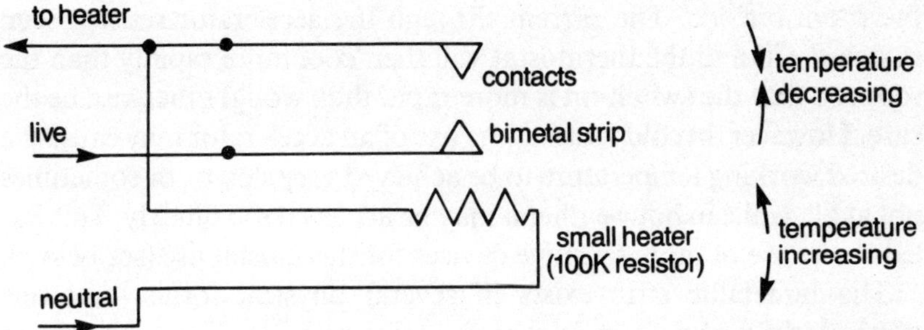

Fig. 4.20 Using a thermostat accelerator. When the contacts are closed, the heater is energised and makes the bimetal strip operate earlier than it would have done in the absence of the accelerator.

same amount, however, the strip will bend as the temperature changes, and this bending action can be used to operate switch contacts, usually with the strip itself carrying one contact.

The conventional type of bimetallic strip element can still be found in some thermostats, though the strip is very often formed into a spiral. This allows for much greater sensitivity, since the sensitivity depends on the length of the strip. Thermostats of this type have an undesirably large hysteresis: a thermostat set for a nominal 20° C might open at 22° C and close again at 18° C. This leads to undesirable temperature swings, causing the occupants of a room to ignore the thermostat and either turn radiators on or off, or use the thermostat itself as nothing more than an on/off switch. The hysteresis of the simple bimetal thermostat can be reduced by the use of an *accelerator*. This consists of a resistor placed close to the element. The principle is that when the thermostat points close to switch on heating in a room, current is passed through the accelerator resistor (Figure 4.20) so that the rate of heating within the thermostat is faster than outside. This leads to the thermostat points opening before the same temperature is achieved in

Fig. 4.21 The disc form of bimetal switch. This is extensively used for over-temperature switches in equipment, and in domestic appliances.

the room outside. The current through the accelerator resistor then switches off, and the thermostat will then cool more rapidly than the room so that the switch-on is more rapid than would otherwise be the case. However, in cold weather the use of an accelerator may cause the desired working temperature to be achieved very slowly, or sometimes not at all, while in hot weather it may be achieved too quickly. This has led to the use of more sensitive devices for thermostat use (see below).

The bimetallic strip exists in several physical forms, and one particularly useful form is the *disc* (Figure 4.21). For a change of temperature, a bimetallic disc will abruptly buckle, giving a snap-over action that requires no form of assistance. This is the basis of the small thermal switches used to guard against overheating in electronic equipment that incorporates heat sinks, small motors, transformers or other components that are likely to overheat and have a metallic surface to which the thermal switch can be bolted. These thermal switches can be bought as normally open or normally closed, depending on whether they are to be used to detect rising or falling temperatures. Since no accelerator is used, there is temperature hysteresis of the order of 3–5 degrees on each side of the preset nominal temperatures. For more precise control, it is possible to obtain units with long bimetal strips that have smaller hysteresis and variable setting temperature. All types of long-element bimetal-strip thermostats should be recalibrated at intervals, since the strip is subject to gradual changes (*creep*) that affect the thermostat setting.

Another principle used in temperature switching is liquid expansion in conjunction with a pressure switch. The sensing element (Figure 4.22) is a capsule filled with liquid that is connected by a narrow-bore tube to the pressure switch. Since the capsule can be remote, and involves no electrical connections, this is often very useful for hazardous environments, and the liquid can be chosen accordingly. The length of connecting tubing must be such that the volume of liquid

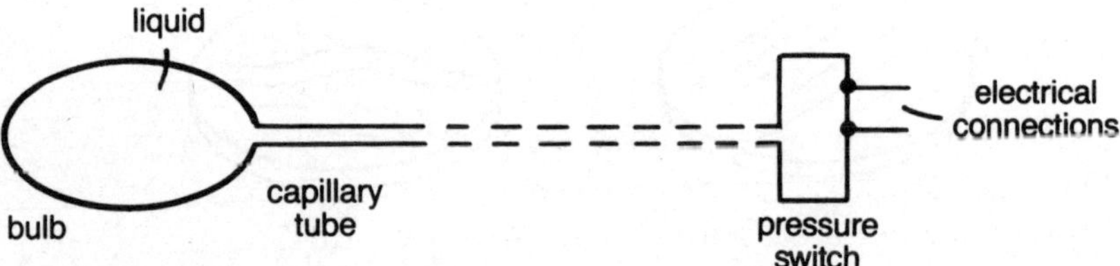

Fig. 4.22 Principle of the liquid-bulb remote switch.

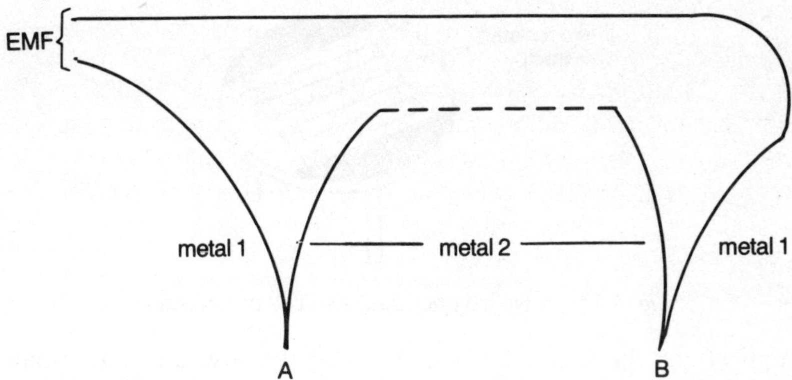

Fig. 4.23 The thermocouple. The voltage generated by such metal combinations as copper and Constantan is small, a few mV for a 10° C difference, but the output can be amplified.

contained in the tubing is only a small fraction of the total volume, since the temperature of the liquid in the tubing will also affect the pressure. The use of air or an inert gas in place of a liquid makes the device much more sensitive, but the pressure switch needs to be able to respond to much lower pressures than are exerted by an expanding liquid.

The *thermocouple* is sometimes used as the sensing element in a thermal switch. The principle is that two dissimilar metals always have a contact potential between them, and this contact potential changes as the temperature changes. The contact potential is not measurable for a single connection (or junction), but when two junctions are in a circuit, with the junctions at different temperatures, then a voltage of a few millivolts can be detected (see Figure 4.23). This voltage will be zero if the junctions are at the same temperature, and will increase as the temperature of one junction is changed, until a peak is reached. Since there is no inherent switching action, the thermocouple must be used along with a *controller*. This will employ a Schmitt trigger type of circuit, and will also permit adjustment of bias so that the switching temperature can be preset. For very high temperatures, the change of resistance of a metal such as platinum can be used in a 'measuring bridge and trigger' type of circuit in place of a thermocouple.For the lower ranges of temperature, electronic sensing methods based on semiconductors are used. The *thermistor* is typical of this type of device, and its change of resistance can be so abrupt that a trigger is often not necessary in the accompanying circuitry. In a very few

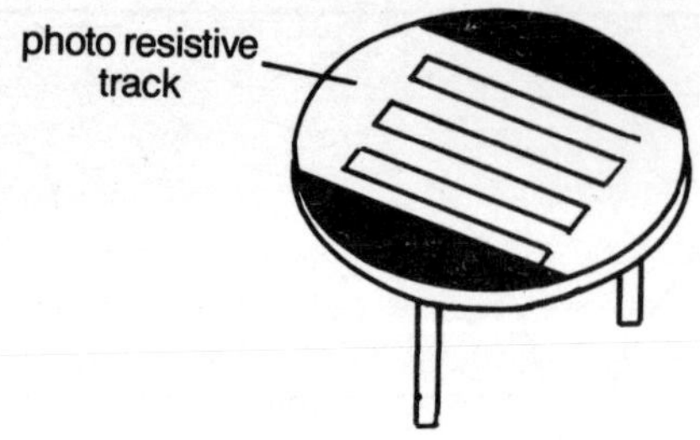

Fig. 4.24 A typical photoresistive cell construction.

applications the thermistor can be used directly, but it is usually undesirable to have the controlled current passing through the thermistor – it is more usual for the thermistor to be part of a transistor switch. This arrangement has the advantage that it can be arranged so as to have zero hysteresis if this is a useful feature. For most switching purposes, however, hysteresis is desirable in order to prevent rapid switching on and off as air currents strike the detector.

Radiant energy switching

Radiant energy, whether in the form of light, heat, radio waves or even ionizing radiation, may be required to operate switches. In most cases the devices used are sensors rather than switches, with the switching action provided by a trigger circuit: a comprehensive description of such sensors belongs elsewhere, and this section will deal with them only briefly. Light-operated switching is dealt with by using *photocells* as sensors, and the most familiar type for industrial use is the cadmium sulphide (CdS) photoresistive cell (Figure 4.24). This cell has a resistance between its terminals that is very high in the dark, but which falls to low values (around 100 Ωohms) in bright light. The cell can be connected between a voltage supply and a relay coil, AC or DC, so that its use does not require amplification, or even the provision of a low-voltage DC supply. By contrast *photodiodes* and *silicon cells* require both amplification and trigger circuits in order to provide a switching output. The detection of radiant heat energy is most efficiently carried out by *pyroelectric* detectors, using materials (mostly ceramic slices or metallised plastic films) whose electrical properties alter when they are struck by radiated heat. These detectors can be encapsulated into TO-5 cans and have replaced the older detectors, such as bolometers.

Radio wave detectors are mainly used for the detection of microwaves in security devices, and make use of diodes in waveguides. Detectors of ionizing radiation use the familiar Geiger-Müller principle in which an ionizing particle or ray passes into a tube filled with gas at low pressure, and momentarily permits a pulse of current to flow. Any detector of the types described will need the addition of a trigger stage, usually in addition to amplification, in order to provide a switching output.

Proximity and level

A *proximity switch* is intended to operate when two objects are close to each other, without mechanical contact. Such a switch must therefore make use of field effects such as magnetism, electrostatics or electromagnetic radiation, and will usually consist of a sensor for whichever effect is used, together with a switching circuit operated by the sensor. The type of sensor that is used will depend on the nature of the material whose proximity is to be detected. For very many purposes, this will be either magnetic material or material into which a magnet can be incorporated, so that a Hall-effect sensor coupled to a transistor-assisted relay can provide proximity action. For lower sensitivity, a magnetic proximity switch can consist of a reed relay, with the object to be detected carrying a magnet. The majority of industrial requirements for proximity switching can be met by these two types of proximity switching. The Hall-effect device has the advantage that its design can include sensitivity controls, so that the proximity distance can be adjusted. When a reed switch is used, adjustment of detection distance is possible only by using a different magnet on the object to be detected.

The magnetic proximity switch is useful only when the object to be detected is magnetic or can carry a magnet, and other methods must be used if this cannot be applied. *Capacitive proximity switching* uses a sensor of changing capacitance coupled to a switching circuit, and is capable of detecting such objects as the human hand, liquid levels and other objects which are not magnetic and/or not metallic. The best form of sensor consists of an oscillator which is tuned by the stray capacitance of a sensing plate (Figure 4.25). The output of this oscillator is fed to a frequency-sensing circuit which in turn operates the switching circuits. When any object approaches the sensing plate, the stray capacitance increases, causing the frequency of the oscillator

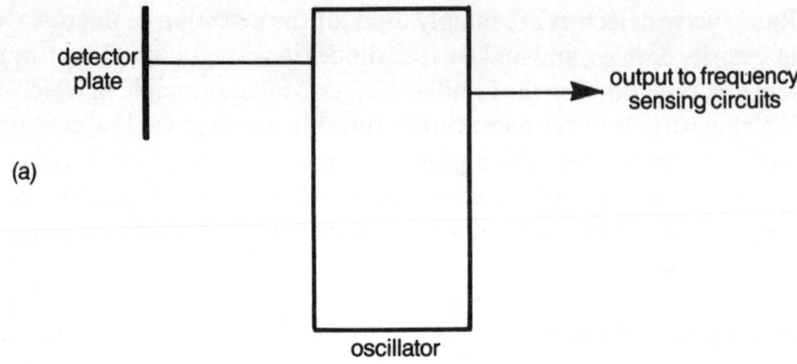

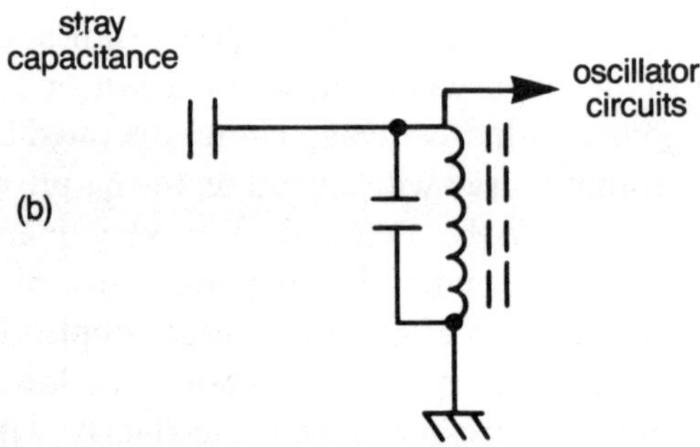

Fig. 4.25 Principle of the capacitance proximity detector. The detector plate forms a capacitor plate of an oscillating circuit whose frequency will be affected by the presence of earthed or isolated material. The frequency change can be detected and used to operate a semiconductor switch or relay.

to decrease. This change is detected, and will cause the switching circuit to operate. Like the Hall-effect magnetic type of proximity detector, this capacitance detector can have its sensitivity adjusted, and its ability to operate on almost any type of material is a considerable advantage.

Of the electromagnetic radiation proximity detectors, the main types are light-beam switches and reflected radio-wave (radar) detectors. The light-beam switch uses a light beam which falls on a photo-detector whose output is used to operate a switching circuit. Interrupting the beam will result in no output from the detector, and so operate the switch. This, however, is not a true proximity detector, because partial interruption of the beam may not operate the switch, and the action takes place only while the object passes through the beam, not before or after. In addition, transparent objects are not detected, so a

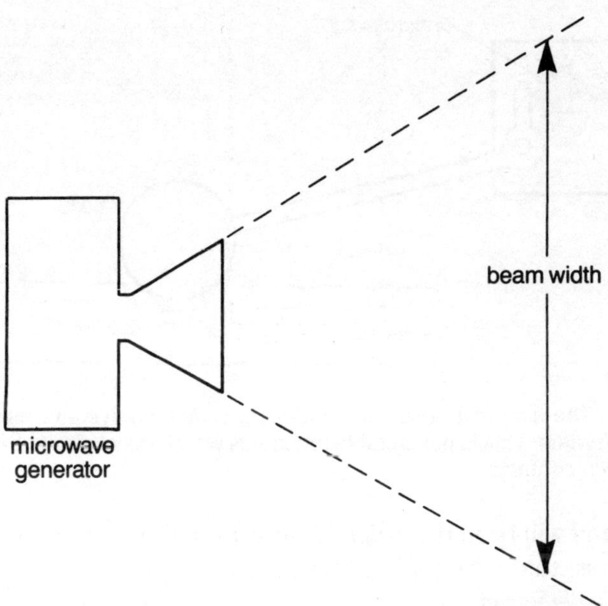

Fig. 4.26 The beam width of a microwave detector system. Any object within the beam will register on the detector, which operates by receiving reflected signals and detecting changes.

light-beam method is valid only when the path of an object can be defined (by, for example, a doorway or hatch) and the type of object and its path is suitable. A light-beam method should never be used if a hazard could arise when an object that has crossed the light beam remains inside the detection zone.

The reflected-beam type of detector is a true proximity detector and can be used with any type of radiation, though radio waves in the radar range are the predominant type. The waves are radiated from an aerial which is usually directional, though the beam angle (Figure 4.26) can be made fairly wide. The beam can be continuous or pulsed, and the presence of any reflecting object in the beam will cause a return of signal. When the beam is continuous, this can be detected on another aerial and the phase of the return beam used to detect motion of the target object. When a pulsed beam is used, the same aerial can serve for detection. The distance of the target is detected in terms of the amplitude of return signal, and the range of detection can be regulated by altering the sensitivity of the amplification of the returned signal. The maximum range is determined by the signal-to-noise ratio of the

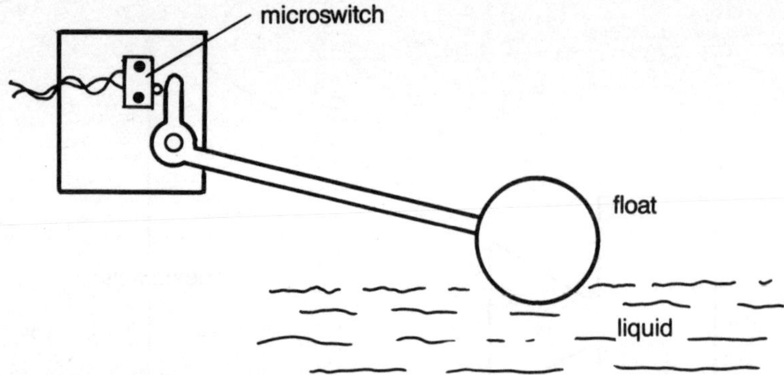

Fig. 4.27 The simplest possible liquid-level switching system, using a float and microswitch. This is not suitable for liquids which may corrode the hinge or microswitch contacts.

system, and will be in the order of yards rather than the millimetres or centimetres that are typical ranges for magnetic or electrostatic proximity detectors.

The reflected beam type of proximity detector can cope with virtually all types of target object, and in some cases a simpler system can be set up using light beams if the object to be detected is light-reflecting, and if the normal illumination is low. In most cases, however, the microwave system is more satisfactory, since it is not disturbed by illumination levels, nor dependent on the target being white or metallic. The microwave system can be obtained ready-packaged so that only the positioning has to be carried out by the user.

Many of the methods employed for proximity switching are also used for liquid level detection. The simplest form of liquid-level switching uses a float and a microswitch, as in Figure 4.27, and for many industrial purposes this is all that is needed. A variation on this system involves fitting the float arm with a magnet that operates a sealed reed switch. This second form has the advantage that the switch contacts are unaffected by the vapour from the liquid, and since the magnet, switch and leads can all be encapsulated this is particularly suitable for corrosive liquids. The switching can make use of low-voltage and low-current supplies if the liquid is flammable. The weak point in all float systems is the hinge of the float arm, which is subject to corrosion if made in metal, or to swelling if plastic.

An alternative that avoids contact with the liquid is to make use of a capacitive detector in a separate tube (Figure 4.28). The tube is fitted

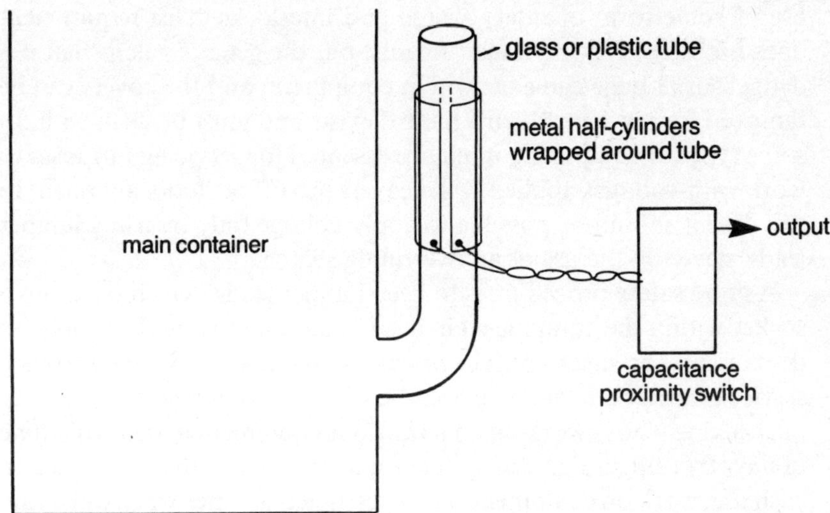

Fig. 4.28 A capacitive liquid-level switch, using capacitance between two curved plates fastened to the outside of a non-metallic tube. The change of capacitance caused by a liquid in the tube between the plates can be made to operate a switch.

with two plates, each wrapped around the tube and not in contact, and the capacitance between these plates is part of the resonant circuit of an oscillator. When the liquid level rises in the tube so that there is liquid between the plates then capacitance will increase, and the resonant frequency of the oscillator will be lowered. This in turn can operate a frequency detector and a switch. The advantage of this type of switching is that nothing need touch the liquid, there are no mechanical actions, and there is no contact between liquid vapour and electrical leads. The main restriction is that the tubing must be non-conducting, usually glass or plastic, which can restrict the temperature range of liquids that can be handled. In addition, if the liquid is liable to clotting or may contain solid particles, there is a risk that the tube may become blocked and that its liquid level will no longer be the same as the level in the main container.

Safety switches and interlocks

Where industrial equipment operates with voltages higher than about 100 V, and contains serviceable parts, all cabinets and cases must make use of some form of safety switch and interlocks. This requirement does not extend to small instruments, but the general rule is that if a dangerous voltage can exist within equipment, and the covers can be removed for servicing, then a hazard exists and must be dealt with. In some types of equipment, it may be essential for servicing purposes to work with voltages applied while covers are off or doors are open. In this case it should be possible to apply voltage only by using jumper leads, never by the use of an overruling switch.

A good safety procedure is to keep jumper leads (which plug into a socket within the equipment) in a safe place, and issue them only on demand and on signature. The practice of issuing jumper leads to each service engineer is a bad one, because when this is done there is no way of checking who is working on hazardous equipment and at what time of day. By using a sign-out system for such leads, with a time limit on each use, work on equipment can be restricted to times when others are at hand to render assistance in case of accident, and failure to return a jumper lead can be used as a warning that an accident may have happened. In addition, jumpers can be regularly tested if they are centrally located.

The most common form of safety switching is the contact switch on each door of a piece of equipment, or the microswitch in the casing of an instrument. These switches usually operate on the mains supply, and must be arranged so that they cannot be manually operated when the door or casing is open. Door switches can usually be manually operated, so the normal practice is to connect several switches in series so as to make it impossible to hold in the plungers of all the switches together. Microswitches used in instrument cases are usually recessed, so that they cannot be finger-operated, though a determined effort with a screwdriver can usually operate the switch. Magnetically-operated safety switches should not be used: it is difficult to check if they have operated, and it is equally difficult to prevent service engineers from carrying magnets that will operate the switches.

Interlocking is another important principle for safety purposes. If, for example, an engineer opens a cage for inspection and steps inside, it should be impossible for dangerous voltages to be switched on if the door should blow shut. This is possible only if the power is applied through a relay which can be operated by way of a push button only when the safety switches are closed. When any of the safety switches opens, the relay is de-energised, and will not be re-energised when the safety switch contacts are made again unless the push button is also

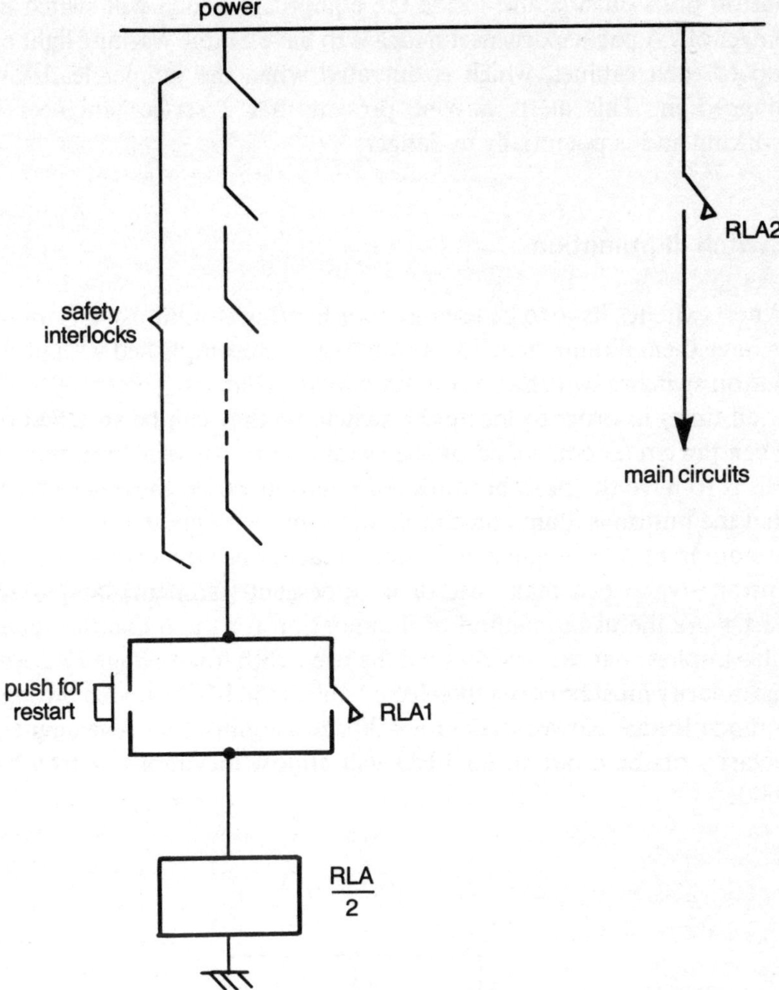

Fig. 4.29 Wiring interlock switches and a master 'on' switch so that closing safety interlocks will not by themselves switch power on.

used. Figure 4.29 illustrates the type of circuit that is used. It should not be possible to operate the push button, or any duplicate switch, from inside the cage or cabinet. If service actions have to be carried out with power on, then this should be done with jumper leads to bridge the safety switches (leaving the door(s) open to make escape or the access of helpers easy) and an assistant should be present to operate the main push button. There should be a large and prominently located push button both outside and inside the equipment which will switch all power off. A particularly useful idea is to have a large warning light on top of each cabinet, which is operated when the jumper leads are plugged in. This alerts anyone present that a service engineer is working and is potentially in danger.

Switch illumination

When switches have to be used in poor lighting conditions, it is useful to have them illuminated. This is most easily accomplished with push-button switches, which can contain miniature lamps. These can be on at all times in order to locate the switch, or they can be switched on when the circuit controlled by the switch is on. A useful variation on this is to have the push button's illumination varied in some way, so that the button is illuminated at all times but with greater intensity or in a different colour when the circuit is made. The larger types of push-button switch can make use of incandescent (filament) lamps, but LEDs are the usual method of illumination for the miniature types. This implies that the switches will be used with low-voltage DC, and the polarity must be correctly selected unless the LEDs are fed through a diode bridge network (Figure 4.30). Remember that reversing the polarity of the diode in an LED will almost inevitably destroy the diode.

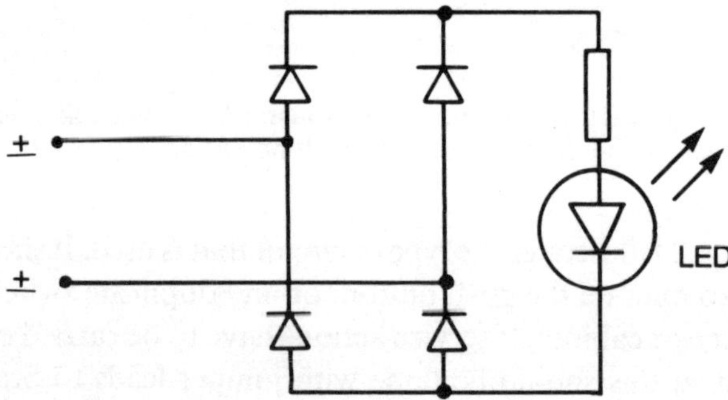

Fig. 4.30 Connecting an LED through a diode bridge so that either AC or DC (or either polarity of DC) can be used.

Non-contact switching

Low-voltage switching can be carried out by semiconductor devices, notably FETs, with no mechanically moving contacts. The use of devices such as FETs and thyristors is more correctly classed as a type of relay action, however, since the switching is in response to an electrical signal. The use of the Hall effect is a true contactless switching effect, because a non-electrical input (the movement of a magnet or a change in a magnetic field) causes a current path to be made or broken. Switching of this type is now very common, notably in the contact-breaker mechanisms of car ignition circuits which use semiconductor switching methods.

Contactless switching has a number of advantages, particularly in hazardous environments. The switching action does not depend on any form of mechanical action, so that problems of rusting, jamming, wear and fatigue no longer exist. Since the 'contacts' are internal to the semiconductor's material they are not liable to corrosion, particularly if the whole switch is encapsulated. Low voltages and currents can be used, avoiding spark risks even if wires are damaged. The switching action can be very fast and in many cases can have variable sensitivity.

The main drawback of contactless switches is that the action generally requires low-voltage DC at some point – I am excluding the action of thyristors and Triacs here. The control action of the switch may, however, need to make or break circuits that operate at higher voltages, and this requires some form of relay action. The conventional contact form of relay can always be used, and in some applications may be required by regulations, but the alternative is a contactless form of relay, the *opto-isolator*. The opto-isolator is shown in outline

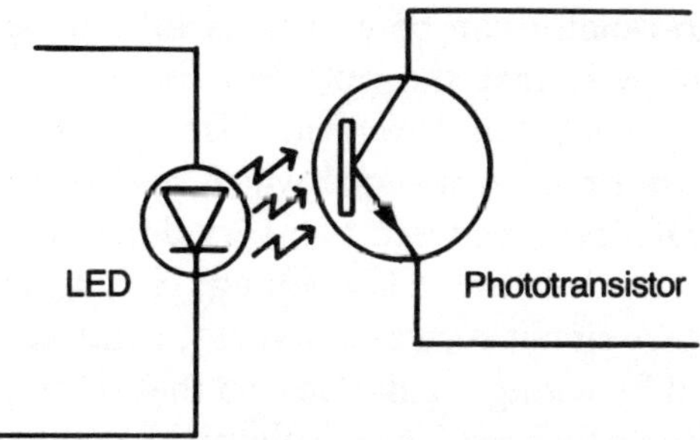

Fig. 4.31 Principle of the opto-isolator, consisting of an LED and phototransistor.

The RS OPL 1264A is a high-voltage opto-isolator which is BASEEFA approved. It consists of a Gallium-Arsenide-Phosphorus emitter coupled to an NPN phototransistor, and provides isolation to 10 kV.

Minimum current transfer ratio (at If = 10 mA, Vce = 5 V)	25%
Rise time	4 μs (typical)
Fall time	3 μs (typical)

Diode:

Vf max at 20 mA and 25 C	1.5 V
Vr min at 25 C	3.0 V
If typical	10 mA
Max. Power dissipation at 25 C	50 mW

Phototransistor:

Vce max	32 V
Max. Power dissipation at 25 C	100 mW

Fig. 4.32 Characteristics of an opto-isolator intended for high-voltage isolation work.

in Figure 4.31. It consists of an LED whose light can be focused onto a photocell, with no electrical contact between the LED and the photocell. When the LED is illuminated, the light path to the photocell ensures that there will be an output from the photocell, and this can be amplified and used for switching purposes. The devices are encapsulated to ensure that all the light from the LED will reach the photocell, and no ambient light can reach the photocell. Figure 4.32 shows some typical opto-isolator characteristics.

Since the LED can be part of a low-voltage circuit, and the photocell can be part of a circuit that is connected to mains or other high-voltage supplies, the opto-isolator can be a very useful linking device. One particular advantage is that the opto-isolator action can be fast, allowing switching at audio and even video frequencies. It is therefore a unique way of transmitting a signal which may contain both high frequencies and DC levels between two very different voltage levels. The most obvious use is to link a low-voltage circuit such as a touch-pad to a high voltage circuit such as a thyristor controller, but this use is often prohibited by wiring regulations, so that some care has to be taken when specifying the use of opto-isolators. Since thyristors can be triggered by pulses applied via transformers, this alternative is more commonly used.

Chapter Five

Specifications and Standards

The two main regulatory bodies in the UK that lay down specifications and standards for the construction and use of switches are the Institution of Electrical Engineers (IEE) and the British Standards Institute (BSI). The BSI are much more concerned with the construction of switches, particularly in respect of standardising dimensions, than with their use, so the BS documents are probably of more interest and relevance to the manufacturers of switches than to the users. The IEE, as might be expected, is more closely concerned with electrical specifications and conditions of use. The following list is of the main IEE publications which deal with the general aspects of switching and wiring, and those which are more specifically directed at switches of the type discussed in this book.

Electrical wiring of all kinds is covered by the IEE regulations, of which the current edition is the fifteenth. This manual of 232 pages covers all aspects of non-military electrical installations, both domestic and industrial, and is now based on the rules of the IEC (International Electrotechnical Commission) and CENELEC (European Committee for Electrotechnical Standardisation). The book is massive, and users will find the appendices of reference data and standard circuits very useful. A separate *Guide* is published which gives more detail, and interpretations of aspects of the general regulations, and this guide has a section on switchgear.

In addition to the general regulations, the IEE publishes a separate book of regulations for the electrical and electronic equipment of ships, and this manual is required reading for anyone concerned with the design or manufacture of such equipment. The relevant legislation is the Merchant Shipping (Passenger Ship Construction) Rules, 1965 and

Statutory Instruments 1103 and 1104, 1965, both made under the Merchant Shipping (Safety Convention) Act of 1948. The regulations now closely follow the guidelines of the IEC publication (No. 92) *Recommendations for Electrical Installations of Ships.* The current edition is the fifth, dated 1972, with amendments in 1974, 1977 and 1982.

Another publication deals with the important case of mobile and offshore equipment. This includes requirements for switchgear, and also for accessories and for communications equipment and navigational aids. This set of regulations dates from 1983.

Regulations for Electrical Installations (15th Edition of the IEE wiring Regulations). 232 pp. 297 × 210 mm, soft covers. ISBN 085296 235 5, 1981 (reprinted 1986). Also available in hardback as ISBN 0 85296 324 6.

Commentary on the 15th Edition of the IEE Wiring Regulations (Revised Second Edition) by B D Jenkins. 288 pp. 219 × 148 paperback. ISBN 0 86341 040 5, 1985 revised edition.

Power Diode and Thyristor Circuits. R M Davis. Deals with semiconductor power switching. 279 pp. 216 × 135 mm paperback. ISBN 0 901223 90 5, 1986 revised edition.

BSI Publications

The BSI publications are more detailed, but are also directed more to the manufacturer than to the user of switches. The following list is of standards which are relevant to switches of various types. The prices which are quoted are for non-members (mid-1987).

BS 9561: 1979 Specification for lever operated switches of assessed quality; generic data and test methods; general rules for preparation of detailed specifications. Price £16.20.

BS 9562: 1982 Specification for microswitches (sensitive switches) of assessed quality; generic data and test methods; general rules for preparation of detailed specifications. Basic and full assessment levels. Price £27.20.

BS 9563: 1979 Specification for rotary (manual) switches of assessed quality; generic data and test methods; general rules for preparation of detailed specifications. Price £16.20.

BS 9563 N001: 1974 (1981) Detail specification for a rotary wafer switch (manual) of assessed quality; 30 mm single hole mounting. Price £8.00.

BS 9563 N0032: 1980 Detail specification for rotary wafer switches of assessed quality; generic data and test methods; general rules for the preparation of detail specifications. Price £16.20.

BS 9564: 1980 Specification for push-button switches of assessed quality; generic data and test methods; general rules for the preparation of detail specifications. Price £16.20.

BS 9565: 1981 Specification for printed board mounted programming switches of assessed quality; generic data and test methods. Price £22.80.

BS 9566: 1984 Rules for the preparation of detail specifications for printed board programming switches. Price £12.20.

BS 9571 N/F000: 1973 Detail specification for lever switches of assessed quality; quick-make quick-break. Price £10.20.

BS 9578 N/F000: 1973 (1981) Detail specification for microswitches (sensitive switches) of assessed quality; 30 mm nominal body dimension, plunger actuated. Price £8.00.

BS CECC 18000: 1984 Harmonized system of quality assessment for electronic components. Generic specification. Dry reed changeover contact units mechanically biased. Price £16.20.

BS CECC 18001: 1984 Specification for harmonized system of quality assessment for electronic components. Blank detail specification. Dry reed changeover contact units mechanically biased, for general application. Price £16.20.

BS CECC 19000: 1979 Specification for harmonized system of quality assessment for electronic components. Generic specification. Dry reed make contact units. Price £22.80.

BS 3676: 1963 Switches for domestic and similar purposes (for fixed or portable mounting). Price £16.20.

BS 4794: Specification for control switches (switching devices) including contactor relays, for control and auxiliary circuits, for voltages up to and including 1000 V AC and 1200 V DC. This consists of the following parts:

Part 1: 1979 General requirements. Price £16.20.

Part 2: Special requirements for specific types of control switches:

Section 2.1: 1977 Push-buttons and related control switches. Price £10.20.

Section 2.2: 1977 Additional requirements for rotary control switches. Price £10.20.

Sections 2.3 & 2.4: 1977 Section 2.3 Contactor relays, Section 2.4 Pilot switches. Price £7.00.

Sections 2.5 & 2.6: 1978 Section 2.5 Indicator lights; Section 2.6 Standardization of fixing hole of single hole mounted push-buttons and indicator lights. Price £10.20.

Section 2.20: 1982 Position switches with positive opening operation. Price £7.00.

BS 5472: 1977 Specification for low voltage switchgear and controlgear for industrial use. Terminal marking and distinctive number. General rules. Price £10.20.

BS 800: 1977 Specification for radio interference limits and measurements for equipment embodying small motors, contacts, control and other devices causing similar interference. Price £16.20.

BS 6134: 1981 Specification for pressure and vacuum switches. Price £10.20.

BS 6517: 1984 Specification for low voltage switchgear and controlgear for industrial use. Single hole mounted control switches and indicator lights. Mounting dimensions. Price £7.00.

BS 6518: 1984 Specification for low voltage switchgear and controlgear for industrial use. Control switches. Position switches 42.5 × 80. Dimensions and characteristics. Price £10.20.

BS 6519: 1984 Specification for low voltage switchgear and controlgear for industrial use. Inductive proximity switches. Identification of connections. Price £7.20.

BS 6520: 1984 Specification for low voltage switchgear and controlgear for industrial use. Position switches 35 × 55. Dimensions and characteristics. Price £10.20.

BS 4727: Part 2: Group 06: 1972 Switchgear and controlgear terminology (including fuse terminology). Price £22.80.

Military Equipment

The Ministry of Defence maintains a standards department in Glasgow, and this deals with the standardisation and specification of all equipment that is bought under MoD contracts. For specialised naval use, reference may have to be made to the MoD department in Bath, Avon. In general, military requirements for electronics components will be for components of assessed quality, and some of the specifications have already been included above. The general guidelines are contained in the BS 9000 series as follows:

BS 9000: General requirements for a system for electronic components of assessed quality:

Part 1: 1981 Specification of basic rules and rules of procedure. Price £10.20.

This, as the title shows, outlines the basic rules and procedures for the assessment of quality of electronic components. The CECC (CENELEC Electronic Components Committee) numbers for specifications are listed (note CECC 16 000 for relays and CECC 19 000 for dry reed relays) and also correlation between BS and CECC numbering. Switches are allocated BS numbers 9560 to 9599. The inspection procedure is noted, along with the point that the BSI issues a qualified products list, PD 9002, which should be noted by purchasing officers. A useful list of organisations and abbreviations is given.

Part 2: 1983 Specification for national implementation of CECC basic rules and rules of procedure. Price £10.20.

This lists the modifications that are needed to make BS 9000 applicable to CECC standards, so that component testing specifications are harmonised throughout the countries concerned.

Part 3: 1987 (Price not available) shows how the IECQ (IEC Quality assessment system) rules of procedure are implemented by BS 9000.

The specifications BS 9563–BS 9578 have been noted previously.

Environmental Testing

The methods of environmental testing are laid down in a set of standards as follows:

BS 2100: Basic environmental testing procedures:

Part 1.1: 1983 General and guidance. Price £16.20.

Part 2.1: Tests. The remaining parts of this specification cover the testing methods for environmental problems as follows:

Part 2.1A: 1977 Cold. Price £16.20.

Part 2.1B: 1977 Dry heat. Price £16.20.

Part 2.1Ca: 1977 Damp heat, steady rate. Price £5.20.

Part 2.1Db: 1981 Damp heat, cyclic (12 hour/12 hour cycle). Price £8.00.

Part 2.1Ea: Shock. 1977 (1984). Price £16.20.

Part 2.1Eb: Bump. 1977 (1984). Price £7.00.

Part 2.1Ec: 1977 Drop and topple. Price £5.20.

Part 2.1Ed: 1977 Free-fall. Price £7.00.

Part 2.1Fc: 1983 Vibration (sinusoidal). Price £22.80.

Part 2.1Fd: 1973 Random vibration – wide band, general requirements. Price £10.20.

Part 2.1Fda: 1973 (1984) Random vibration – wide band. Reproducibility high. Price £16.20.

Part 2.1Fdb: 1973 (1984) Random vibration – wide band. Reproducibility medium. Price £12.20.

Part 2.1Fdc: 1973 (1984) Random vibration – wide band. Reproducibility low. Price £8.00.

Part 2.1Ga: 1984 Acceleration, steady state. Price £8.00.

Part 2.1J: 1977 Mould growth. Price £7.00.

Part 2.1Ka: 1977 Salt mist. Price £7.00.

Part 2.1Kb: 1977 Salt mist. Price £5.20.

Part 2.1Kc: 1977 Sulphur dioxide test for contacts and connections. Price £8.00.

Part 2.1Kd: 1977 Hydrogen sulphide test for contacts and connections. Price £7.00.

Part 2.1M: 1984 Low air pressure. Price £7.00.

Part 2.1N: 1985 Change of temperature. Price £12.20.

Part 2.1PZ: 1970 Flammability. Price £5.20.

Part 2.1Q: 1981 Sealing. Price £16.20.

Part 2.1R: 1976 Resistance to fluids. Price £8.00.

Part 2.1Sa: 1977 Simulated solar radiation at ground level. Price £7.00.

Part 2.1T: 1981 Soldering. Price £16.20.

Part 2.1U: 1984 Robustness of terminations and integral mounting devices. Price £16.20.

Part 2.1XA: 1981 Test and guidance: Immersion in cleaning solvents. Price £8.00.

Part 2.1Z/AD: 1977 Composite temperature/humidity cyclic test. Price £8.00.

Part 2.1Z/AFc: 1984 Combined cold/vibration (sinusoidal) tests for both heat-dissipating and non-heat-dissipating specimens. Price £10.20.

Part 2.1Z/AM: 1977 Combined cold/low air pressure tests. Price £8.00.

Part 2.1Z/AMD: 1977 Combined sequential cold, low air pressure and damp heat test. Price £5.20.

Part 2.1Z/ABDM: 1983 Climactic sequence primarily intended for components. Price £5.20.

Part 2.1Z/Bfc: 1984 Combined dry heat/vibration (sinusoidal) tests for both heat-dissipating and non-heat-dissipating specimens. Price £10.20.

Part 2.1Z/BM: 1977 Combined dry heat/low air pressure tests. Price £8.00.

Part 2.2 Guidance.

Part 2.2C and D: 1981 Guidance, tests C and D. Guidance for damp heat tests. Price £8.00.

Part 2.2J: 1977 Mould growth. Price £7.20.

Part 2.2Kd: 1984 Hydrogen sulphide test for contacts and connections. Price £8.00.

Part 2.2N: 1977 Guidance on change of temperature tests. Price £7.00.

Part 2.2Sa: 1977 Guidance for solar radiation testing. Price £12.20.

Part 2.2T: 1981 Guidance on soldering test. Price £7.00.

Part 3: Background information.

Parts 3A & B: 1977 Tests A (cold) and B (dry heat). Price £16.20.

Parts 3A & B: Supplement No. 1: 1980 Tests A (cold) and B (dry heat). Price £5.20.

Parts 3Z/AM & Z/BM: 1977 Combined temperature/low air pressure tests. Price £7.00.

Part 4: Section 4.1: 1983 Specification for mounting of components, equipment and other articles for dynamic tests. Price £10.20.

Part 4.2: 1984 Guidance on the application of the tests of BS 2011 to simulate the effects of storage. Price £7.00.

BS 5490: 1977 Specification for degrees of protection provided by enclosures. Price £16.20.

BS 6448: Fire hazard testing for electrotechnical products.

BS 6448 Part 2: Section 2.2: 1984 Needle-flame test. Price £12.20.

PD 6484: 1979 (1984) Commentary on corrosion at bimetallic contacts and its alleviation. Price £16.20.

Index

AC leakage path, 11
accelerated test, 21
accelerator, thermostat, 77
actuating mechanism strength, 21
actuation, microswitch, 31
air actuators, 60
air, ionization, 5
alternate action, 24
alternate action pb switch, 29
anode of thyristor, 71
arc suppression, 7, 8
arcing, 5

biased switch, 23
bimetallic disc, 78
bimetallic switch, 76
bipolar transistor switch, 46
bounce, switch, 23
bridge circuit, 46
broad-band noise, 39
BSI, 91
burning, 2
bush mounting, 24

cadmium sulphide cell, 80
cam mechanism, 57
capacitance, 35
capacitance, switch, 34
capacitive level detector, 84
capacitive proximity, 81
capacitor for suppression, 11
cathode of thyristor, 71
CENELEC, 91
changeover, 23
chopper switch, 47
coaxial construction, 56
cold side earthing, 37
connections to switches, 15
connectors, 15
contact bounce, 49
contact burning, 2
contact crater, 7
contact dissipation, 18
contact materials compared, 4
contact mound, 7
contact potential, 39
contact resistance, 2, 18, 20
contact resistance, signal switch, 33
contact sticking, 3
contactless switchings, 89
controller, thermocouple, 79
crater on contact, 7
creep effect, 78
crimped connections, 15
crosstalk, 37
current type, 5

Darlington pair circuit, 64
debouncing switch, 70
decoded keyboard, 55
detent for rotary switch, 40
differential, microswitch, 30
digital switching, 48
DIL switch, 28, 52

diode suppression, 9
dissipation at contacts, 18
door switch, 86
dummy wafer, 42
dynamic pressure, 23

earthing cold side, 37
effects of temperature, 14
electrical life, 21
electroplating, 3
environmental testing, 96
expansivity, 76

FET switch, 47
filament lights, 88
films of oxide, 2
flameproofing, 14
foot operation, 60
four layer device, 71
freezing of lubricant, 14

gate of thyristor, 71
gate turn-off (GTO), 75
gold plating, 3
GTO, 75

Hall effect, 64
Hall voltage, 67
high currents, 18
high sensitivity relay, 64
hold in, relay, 62
hysteresis of temperature, 77
hysteresis, mechanical, 18

IEC, 91
IEE, 91
ignition points, 2
impulse noise, 39
inductive load, 6
inert gas filling, 15
inhibit pin, digital, 49
intelligent keyboard, 55
interlocks, 86
interrupt driven keyboard, 52
ionizing air, 5
iridium, 4

joystick action, 58
jumper leads, 86

key action of switch, 30
key lock, wafer, 42
keypad, 53
knife switch, 22

latching relay, 64
latching switch, 29
leakage path, AC, 11
LED lights, 88
lever mechanism, 58
light-beam switch, 82
light-operated switch, 80
liquid expansion switch, 78
liquid level switch, 84
local overheating, 19
low-bounce, action, 43
low-friction plastics, 14
low-torque microswitch, 31
lubricant, freezing, 14

magnetic operation, 61
magnetic proximity, 81
mains switch, wafer, 42
manual operation, 21
matrix keyboard, 53
mechanical hysteresis, 18
mechanical life, 21
Merchant Shipping Act, 91
mercury tilt switch, 75
mercury vapour hazard, 76
mercury-wetted reed, 43
metal transfer, 6
microswitch, 30
microswitch actuation, 31
military equipment, 95
molybdenum, 5
momentary action, 24
momentary action pb switch, 29

mound on contact, 7

neutral position, 27
nickel-silver, 5
nitrogen, 15
noble metals, 4
noise, electrical, 39
non-contact switching, 89
non-linear resistors, 13

opto-isolator, 74, 89
oxidation, 3
oxide films, 2

paddle-shaped toggle, 26
palladium-silver, 5
patch in software, 52
pcb mounting, 26
photocells, 80
photodiode, 80
photoresistive cell, 80
piggy-back receptacles, 16
plasma, 7
plate mounting, 25
platinum, 4
pneumatic operation, 60
poles, 23
polled keyboard, 52
proximity switch, 81
pull in, relay, 62
push-button switch, 29
pyroelectric detectors, 80

radar detector, 83
radiant energy switching, 80
radio frequency interference (RFI), 12
rectifying contact, 34
reed switch, 42
reflected beam detector, 83
relay action, 61
reliability, 21
remanence relay, 64
resonance, 10
resonant frequency, 11
RFI, 12
rhodium, 4
rocker switch, 26
rotor, wafer switch, 40
rotary mains switch, 30
RS flip-flop, 49

safety switch, 86
Schmitt IC, 70
Schmitt trigger, 67
Schottky diodes, 56
SCS, 75
sealing glands, 26
sealing materials, 15
secondary switch, 42
side switch, 28
signal noise, 38
silicon cells, 80
silicon-controlled switch (SCS), 75
silicone rubber, 15
silver-coating, 3
silver contacts, 3
smart keyboard, 55
snap-over mechanism, 17
soldered connections, 15
solid-state switch, 8
solid-state switching, 45
sparking, 5
specifications, 91
speed of action, 35
standards, 91
sticking of contacts, 3
stray capacitance, 37
sulphur dioxide action, 3
suppression by capacitor, 11
suppression by diode, 9
switch bounce, 23, 67
switch connections, 15
switch contacts, 2
switch illumination, 88
switch size, 20
switching joystick, 58
synchronous switching, 48

temperature effects, 13
thermal safety switch, 78
thermal switch, 76
thermistor, 79
thermocouple, 79
thermostat accelerator, 77
throws, 23
thyristor, 71
thyristor circuits, 73
toggle, 22
toggle switch, 24
toggling action, microswitch, 30
transfer of metal, 6
trapped key, 30
Triac, 74
tungsten, 5
turn-off, thyristor, 73
type of current, 5

UHF signal switching, 56

VHF signal switching, 56
voltage hysteresis, 71

wafer rotary switch, 29
wafer switch, 37, 40
wafer-type joysticks, 58
waterproofing, 15
ways, 23
welded connections, 15
wiping action, 2